POSTHUMAN SOCIAL SCIE COMPUTATIONAL CULTUR

Posthuman Social Science and Computational Culture offers theoretical and practical insight into posthuman social science research methods, addressing new challenges in computational culture. Philosophy and theory from media studies, software studies, and science and technology studies (STS) are used to make sense of cultural trends and empirical experiments.

This book explores new empiricisms in posthuman social science, with particular attention to computational culture and digital life. This book carefully treats key philosophical thinkers such as Karen Barad, Rosie Braidotti, Gilles Deleuze, Édouard Glissant, Bruno Latour, and Gilbert Simondon, and explains how these theorists help make sense of our current social and material conditions. Essays explore difficult questions for contemporary social science researchers: How should we understand computability and its limits? What is the nature of digital relationality? In what sense are software instruments useful in studying political ecologies? How do sensory and affective atmospheres become robust learning environments? When do arts-based and surrealist experiments have impact? Can mapping experiments help disrupt oppressive spatial regimes where control dominates? Why might we affirm the speculative practices of both science and fiction, as part of a post-colonial Anthropocene epistemology? What differentiates human and machine cognitive behavior? Ideas from many current media theorists in software and cultural studies are taken up and applied, including the work of Louise Amoore, Matthew Fuller, Alexander Galloway, Mark Hansen, Kara Keeling, Achille Mbembe, and Luciana Parisi.

This book will be of interest to graduate students and researchers in the social sciences and humanities (education, social work, sociology, anthropology, media studies, feminist studies), who are interested in the power of theory and philosophy, and are designing empirical experiments to investigate the complex ways that computation and digital technology are changing our lives and our sense of how and why we do "social" research.

Elizabeth de Freitas is Professor and Co-Founder of MIXI, Adelphi University's Manhattan Institute for Studies of STEM and the Imagination, USA. She works across disciplines, developing creative and critical research methodologies for the social sciences, including mapping, and other visual methodologies. With expertise in Science and Technology Studies, and the history and philosophy of mathematics, her research explores the power of speculative thought in social inquiry, and the role of algorithmic thinking in control societies. Her work has been funded by the Canada Council for the Arts, the Toronto Arts Council, the Social Sciences and Humanities Research Council of Canada, the US National Science Foundation, the UK Economic and Social Research Council, and the Spencer Foundation.

Matthew Curinga is Associate Professor at Adelphi University and Co-Founder of the MIXI Institute for STEM and the Imagination. His research is concerned with the ways that software design and technical systems influence social and political life, with a focus on the impact of computer science and computational practices on shaping education. His current focus on geospatial systems and cartographies highlights the entanglement between digital and material spaces. His research has been funded by the Mozilla Foundation, the National Science Foundation, and the Spencer Foundation.

Maggie MacLure is Professor Emerita, Manchester Metropolitan University, UK and a former member of the Education and Social Research Institute, where she led the Theory and Methodology Research Group. She is the Founding Director of the International Summer Institute in Qualitative Research. She has directed major research projects and published extensively in her areas of research expertise, which include the development of theory and methodology in applied social research and innovative approaches to the study of early childhood.

David Rousell lives and works on the unceded lands of the Eastern Kulin Nation in Naarm (Melbourne), Australia. He is Associate Professor of Regenerative Education at RMIT University, where he teaches and researches in the areas of climate justice, environmental arts and design, decolonial studies, and regenerative futures. He is a core member of the Centre for Urban Research, co-convener of the Critical Forest Studies Collaboratory, and fellow of the Regenerative Futures Institute.

Laura Trafí-Prats is a Reader in the School of Education at Manchester Metropolitan University. Her research is situated in the field of creative education and grounded in feminist new materialisms, theories of affect, and visual culture studies. She explores how young people experience urban environments using sensory ethnographic methods, such as video, LiDAR, mapping, and modeling. She was co-lead on the project *Mapping spatial practices and social distancing in smart schools: Sensory and digital ethnographic methods*. She was co-editor of *Visual Participatory Arts-Based Research in the City: Ontology, Aesthetics and Ethics,* published by Routledge.

Sarah Truman is a transdisciplinary scholar and Associate Professor at the University of Melbourne. As an ARC DECRA fellow (2022–2025), Truman led *Speculative Futures*, a project exploring speculative fiction as a method for interdisciplinary teaching and research. Author of *Feminist Speculations and the Practice of Research-Creation* (2022) and co-author of *Walking Methodologies in a More-than-Human World* (2017), Truman's prior and ongoing collaborations include co-directing the Literary Education Lab, the international arts collective *WalkingLab*, and the electro-folk group Oblique Curiosities.

POSTHUMAN SOCIAL SCIENCE AND COMPUTATIONAL CULTURE

Essays on Methodology, Theory, and Practice

Elizabeth de Freitas
With Matthew Curinga, Maggie Maclure, David Rousell, Laura Trafí-Prats and Sarah Truman

Routledge
Taylor & Francis Group
LONDON AND NEW YORK

First published 2026
by Routledge
4 Park Square, Milton Park, Abingdon, Oxon OX14 4RN

and by Routledge
605 Third Avenue, New York, NY 10158

Routledge is an imprint of the Taylor & Francis Group, an informa business

British Library Cataloguing-in-Publication Data
A catalogue record for this book is available from the British Library

ISBN: 978-1-032-88627-5 (hbk)
ISBN: 978-1-032-87720-4 (pbk)
ISBN: 978-1-003-53876-9 (ebk)

DOI: 10.4324/9781003538769

Typeset in Optima
by codeMantra

CONTENTS

ACKNOWLEDGMENTS

I would like to first thank all the collaborators, co-authors, editors, and research assistants who have supported this research. In particular, I want to thank my co-authors who made these essays into fabulous adventures of learning and speculation: Matthew Curinga, Maggie Maclure, David Rousell, Laura Trafi-Prats, and Sarah Truman. I would also like to thank Harry Torrance for his wonderful mentorship and leadership at the Education and Social Research Institute at Manchester Metropolitan University. None of this could have happened without Harry! My gratitude goes to the many editors who have toiled away and given so many hours to help shape these chapters as they emerged. In particular, Vivienne Bozalek, Sarah Diem, Kalervo Gulson, Dan Harris, Rachel Holmes, Alecia Jackson, Débora Lanzeni, Hillevi Lenz-Taguchi, Lisa Mazzei, Bernadette Peters, Sarah Pink, Rachel Smith, Bettie St. Pierre, Karen Waltorp, and Michelle Young. I would also like to thank colleagues and friends whose work has been an inspiration and whose support through various academic networks was crucial for the writing of these essays: Felicity Colman, Ezekiel Dixon-Román, Anna Hickey-Moody, Patti Lather, Aislinn O'Donnell, Helen Palmer, Nathalie Sinclair, and Iris Van der Tuin.

This is a tumultuous time for research methodology across both the social and physical sciences, as computational practices and digital tools change research practice, and as the terms social, material, truth and facticity are renegotiated. This book responds to this challenge. Each chapter engages with this tumult and sets a course for inquiry in the havoc of the post-digital age. These essays are rich with theory and empirical practice, offering ideas, tools, and strategies for social science researchers across disciplines. My hope is that scholars new and old, and across the humanities and social sciences, will find relevance in these essays, as they grapple with shifting academic and intellectual terrains and the disintegration of standard images of science and inquiry. This book consolidates my work on social science methodology over the last decade, where I have sought to build models of inquiry that respond to the new onto-epistemological conditions of life and learning, as part of a new stage in the relentless commodification of life through the algorithmic capture of data. The social sciences are poised to not only serve but also contest and resist burgeoning methods of data capture, especially those that claim to map subconscious dimensions of behavior, using predictive data profiling of individuals. My hope is that this book not only is timely in offering readers a much-needed adventure in posthuman social science, but also offers readers a rigorous treatment and elaboration of theoretical perspectives and some epistemological paths forward.

As I publish widely across some far-flung books and journals, this compilation is a chance to bring these together, and, by doing so, raise new insights about digital imperialisms and the incomputable. I've selected material that speaks directly to our current theoretical and methodological challenges in the social sciences. This book includes adaptations of the following journal articles and book chapters. Chapter 2 is based on the journal article "Science studies and the multiple earth: Bruno Latour's risky diplomacy," published in 2020, in *Cultural Studies – Critical Methodologies*, 20(3), 203–212. Chapter 3 is based on the article "The new empiricism of the fractal fold: Rethinking monadology in digital times," published in 2017, in *Cultural Studies – Critical Methodologies*, 16(2), 224–234. Chapter 4 is based on an article published in 2017, called "Karen Barad's quantum ontology and posthuman ethics: Rethinking the concept of relationality." *Qualitative Inquiry*, 23(9), 741–748. Chapter 5 is based on an article "Classroom video data and the time-image: An-archiving the student body," published in 2015, in *Deleuze Studies*, 9(3), 318–336. Chapter 6 is based on an article "The biosocial subject: Sensor technology and worldly sensibility," published in 2018, in *Discourse: Studies in the Cultural Politics of Education*, 39(2), 292–308. Chapter 7 is based on the book chapter, "Eco-sensor technologies and the surrealist impulse," published in 2023, in *An Anthropology of Futures and Technologies* edited by Débora Lanzeni, Karen Waltorp, Sarah Pink, and Rachel C. Smith. Chapter 8 is based on a book chapter, "Atmospheric data and software arts: New ways of investigating the built environment" co-authored with Laura Trafi-Prats, and published in 2023, in *Postfoundational Approaches to Qualitative Inquiry*, edited by Lisa A. Mazzei and Alecia Y. Jackson. Chapter 9 is based on the journal article "New possibilities for spatial research: Re-animating the environment through speculative maps and models," co-authored with Laura Trafi-Prats, and published in 2024 in *Qualitative Inquiry*. These last two chapters were the outcome of a research project, *Mapping spatial practices and social distancing in smart schools: Sensory and digital ethnographic methods*. ES/V006436/1, funded by the UK Economic and Social Research Council as part of the Transformative Methods Call 2021–2023. The project's open access data can be consulted at Mapping Spatial Practices and Social Distancing in Smart Schools: Sensory and Digital Ethnographic Methods. [Data Set]. MMU e-space. DOI: 10.23634/MMU.00631426. Chapter 10 is based on the journal article "New empiricisms in the Anthropocene: Thinking with speculative fiction about the future of social inquiry," published with co-author Sarah Truman in 2020, in *Qualitative Inquiry*, 27(5), 522–533. Chapter 11 is based on my keynote at the New Materialism Network Meeting in South Africa, and an associated article published in 2020, in *Cristal: Critical Studies in Teaching and Learning*. Chapter 12 is based on research with my colleague at Adelphi University, Matt Curinga, where we have been investigating algorithmic intrusions into learning theories, and studying big data patterns in NYC school open portal data (see https://mixi.nyc/ for more projects).

Chapter 13 is based on the chapter "The spatial logic of epistemic imaginaries: Guided by Édouard Glissant" published in 2023 in the *Handbook of Critical Education Research: Qualitative, Quantitative and Emerging Practices*, edited by Michelle D. Young and Sarah Diem. Finally, Chapter 14 is based on an article published in 2024, called "Fragile texts and machine readers: Trans/in/dividual reading tactics in a complex technical milieu," in the *International Journal of Qualitative Studies in Education*, 37(6), 1655–1665.

1

INTRODUCTION

Where lies the incomputable?

The cosmic artisan

This book is a compilation of essays focused on posthuman social science research methodology, theory, and practice. Much of this work was sparked while I was co-director with Maggie Maclure of the Manifold Laboratory for Studies of Learning and Behavior at Manchester Metropolitan University. The lab was created to be a hub for interdisciplinary research across the humanities and social-physical sciences, to think with theory about digital life. The lab was formed at a key historical moment when social science researchers were becoming increasingly aware that computational cultures demanded new kinds of qualitative and quantitative methods of inquiry. This coincided with widespread concerns about the climate crisis and the need to re-imagine science-and-society in light of the Anthropocene and Capitalocene (Hird & Yusoff, 2019; Latour, 2017, 2018).

My aim in composing this book is to offer scholars a kind of foundation for posthuman practices of empirical inquiry. The name "posthumanism" is not as important as the substance, and I ask that the reader not dither too much over the brand, especially since I am more inclined to use other terms that have less currency and offer a more playful absurdity and baroque audacity: for instance, I see the "cosmic artisan" in all these chapters. The term "cosmic artisan" was coined by Deleuze and Guattari (1987) to characterize material craft-like artisanry that was both responsive to local conditions and of worldly significance because it operated through creative abstraction and philosophical inquiry. The cosmic artisan taps into the imperceptible forces that animate the earth, playing artfully with conceptual matter, and "developing an ethico-aesthetic paradigm of ecosophy" (Jagodinski, 2023). A cosmic artisan unleashes the virtual and diagrammatic possibilities of a particular problem and its associated milieu, making sensible what was previously imperceptible (Sholtz, 2021). The figure of the artisan is contrasted to both the artist and the industrialist, and conjures a very specific practice and relation to the material-technical milieu. As the philosopher of technology Gilbert Simondon (1958/2017) suggests, the artisan follows the matter-flows and moves with the indeterminacy of processual matter. The carpenter, for instance, follows the grain of the wood, the metallurgist follows the strain of

DOI: 10.4324/9781003538769-1

the alloy, the brick maker follows the imperfections of the brick-mold. In a world of flux and e-flutter, where digital life supports a new digital oceanic wave of affect, the artisan must be both follower of molecular flow and prospector of the virtual. This is a different kind of prospecting from the data mining and lithium extraction pursued by technocratic industry, because the artisan cannot risk the destruction of the milieu. Their livelihood depends on it. They immerse themselves in the milieu, and become altered as they attune to the flow they follow. The practice of the cosmic artisan must include philosophical investigation. Such a perspective helps characterize my approach to research methods in computational cultures. Cosmic artisans cannot work with ready-made software applications as prescribed, but must become the itinerant or ambulant investigator, reassembling their equipment in response to changes in the milieu and designing experiments based on philosophical investigation.

State science tends to tame the wild investigative practices associated with the artisan, stripping away the materiality of their inquiry methods, and repackaging them as molar state-sanctioned procedures for marshalling and warranting truth. Creative approaches to social inquiry often become simplistic practices within Academia as they are legitimized and made actionable. Simondon (1958/2017), whose work informs many of my claims about technology throughout these essays, will contrast the artisan to the industrialist, but reverse the usual ways in which these two are contrasted and argue that the artisan works on a *more abstract* plane. While the industrialist concretizes and specializes and traps methods into closed machinic loops, the artisan must stay with the open machine, the pluri-functionality of technics, and the indeterminacy of the process. Thus, the artisan operates with a kind of abstract technics *precisely because* the artisan is localized and responsive to the milieu. The artisan researcher engages materiality with an embodied practice, but this practice is alive with the potentiality of the immaterial, and the indeterminacy of the virtual. And the artisan sustains this encounter, again and again, because the virtual potential of materiality is itself flexible and decomposable, unlike the rigid ossified practices of the industrial paradigm.

The figure of the cosmic artisan helps us articulate how we might develop a research practice that is both idiosyncratic and cosmic, singular and yet worlding. This is a practice that is "minor" because its power must come, at this historical juncture, from outside the mainstream or major traditions of inquiry. Rather than striating the social, as imperial science does, the artisan seeks out the subsoil and the metamorphic zone, not to extract its power, but to follow its flow and undulation with an attunement to the earth or cosmos as a whole. The artisan follows the nomadic movement of force and affect in any given environment, so that *following* becomes a method of inquiry, a "following" that is attuned to the larger cosmos, or what Glissant (1999) called "Relation" and "tout-monde." Following is also an extremely dangerous practice under algorithmic conditions where digital connectivity makes following a mere "click" without adequate understanding of the consequences of such actions. In the midst of this haphazard digital "following," the demand for new social science methodologies is all the more urgent. What new rendering of "the social" is emerging in this rapidly changing milieu? How might a social science help us address the fragmentation of online associations, and the isolation of alternative truths, so that the "patchy Anthropocene" (Tsing et al., 2023) might be gathered together into a collective effort with the earth? The Caribbean philosopher and poet Édouard Glissant (1997,

1999) speaks of "world imaginaries" and "world orientation" as well as key terms such as "tout-monde," "écho-monde," and "chaos monde" and "earth aesthetics." Glissant was a poet and philosopher situated historically on the island of Martinique, which allowed him to generate a powerful theory of creolization and post-colonial futurity; he resisted the logic of center/periphery or empire/colony in all his work, always stretching nomadically out beyond the local, gathering the world as a whole or *tout-monde*, enlivened through multiplicity and fluid or metamorphic Relationality. This cosmic worlding gesture, so dangerous and yet so necessary, formulated from the archipelagic scattering and ", diffractive" Caribbean, makes Glissant an extremely important thinker for our times.

Many of these essays are collaboratively written and bring together theoretical perspectives with empirical experiments in the spirit of the cosmic artisan to show how theory and experiment work together. The theory stays tuned to a particular cluster of philosophical ideas, influenced largely by my reading of Gilles Deleuze, Bruno Latour, Gilbert Simondon, and Édouard Glissant. Their ideas are essential across these chapters, in their explicit engagement with the limits of classical empiricism and their attempts to propose alternative practices that might better engage with post-colonial and post-cybernetic onto-epistemic mixtures. Deleuze (1994) pursued transcendental empiricism as a philosophical break from the long legacy of Kant; Simondon formulated alternative human-technology relationships as a supplement to Marx; while Latour studied "down to earth" metamorphic zones as a sociological break with Bourdieu; and Glissant sought new part/whole relationality in the post-colonial archipelagic "tout monde" of the Caribbean.

Of equal significance in this work are feminist thinkers Karen Barad and Rosi Braidotti, who figure prominently in these chapters and whose work is crucial for rethinking social science. Barad has forged new ways of understanding the onto-epistemology of post-Newtonian science, while Braidotti has formulated a specific feminist approach to posthumanist inquiry (Braidotti & Bignall, 2019). Other feminist thinkers show up in these essays, to help pursue this agenda: Elizabeth Grosz, Donna Haraway, Kara Keeling, Elizabeth Wilson, Sylvia Wynter, and more. Braidotti embraced the term posthumanism to characterize a new criticality (Braidotti, 2021), linking it to an ethics of care and cross-species relationality, drawing from Spinozist ontology and Haraway's cyborg feminism. According to Braidotti (2021), posthumanist feminism rethinks life "as a generative force beneath, below and beyond what we have made of it" (p. 240). This kind of vitalism is often associated with indigenous philosophies and new materialisms, and Braidotti traces the ways in which these ideas have informed "postcolonial ecocriticism" and a "zoe-centered egalitarianism" which "draws productive parallels between the species, while preserving the distinct capacities, propensities, and predispositions of humans" (p. 84). Her use of the suffix zoe and geo, rather than bio, is explicitly meant to move the focus toward modes of life that are more vulnerable, monstrous, and "deviant" from humans (p. 219).

Throughout this book, there is a sense that this kind of research produces a "cabinet of curiosities" in the baroque tradition, making visible and manifest the strange new configurations of digital life. My hope is that each essay offers a distinctive dramatization of what is all around us and yet barely noticed, namely our more-than-human computational culture. I use the term baroque here to designate a distinctive way of investigating the world that affirms the power of speculation in coming to knowledge (Chen-Morris,

2022). Baroque thinkers such as Athanasius Kircher (1602–1680), Margaret Cavendish (1623–1673), and Sor Juana Inez de la Cruz (1651–1695), sought a methodology that was grounded in observation, data, art, and invention. They wrote science fiction and poetry and ecstatic dream-like journeys as a way of doing onto-ethical-epistemology. But they also climbed into volcanoes to study vulcanism, collected tidal data from all over the Earth, wrote polemics for indigenous and feminist equity, and believed that experiential epistemic warranting was crucial. They were cosmic artisans. They were part of a Baroque science and culture that was learning new ways to think with instruments and technology (the telescope, the microscope, the camera obscura, the sunflower clock, …), and they had to rethink both the role of human senses and the multiplicity of new perspectives in knowledge production.

The baroque is tapped by both Deleuze and Glissant in similar ways to un/fold the current horizon of research practice; the baroque becomes a way of composing new heterogeneous and yet syncretic worlds, a practice that can emerge before and after the 17th century. As elaborated in Chapter 14, Glissant (1997) describes "creolization" as a form of "baroque derangement" of colonial European classicism (Glissant, 1997, pp. 91–127). He presents a list of possible candidates for such a practice:

> William Faulkner's work, Bob Marley's song, the theories of Benoit Mandelbrot, are all échos-monde. Wilfred Lam's painting (flowing together) or that of Roberta Matta (tearing apart); the architecture of Chicago and just as easily the shanty towns of Rio or Caracas; Ezra Pound's Cantos but also the marching of schoolchildren in Soweto are échos-monde.
>
> *(Glissant, 1997, p. 93)*

The baroque entails "vertiginous variance" (Glissant, 1997, p. 102) and vitalism, an awareness that measurement is always relational, analogical, and processual (de Freitas & Sinclair, 2022). This is by no means an innocent activity, as the baroque was precisely how the European colonial aesthetic folded itself into other peoples' territory, and how capital proliferated through combinatoric variance and analogical mixing. Thus, I use the term baroque here because of its ambivalence, because it reminds me that the imagination, speculation, fantasy, and fabulation are always dangerous, though also necessary, for effecting change.

Because many of the essays are explicitly engaged with the computational turn, thinkers from media theory and software studies also figure prominently. I have tried to respond to the huge changes in the humanities and social sciences over the last decade, as software and AI have come to saturate both qualitative and quantitative research methodologies. The insights of media theorists, such as Matthew Fuller, Alexander Galloway, Mark Hansen, and Luciana Parisi, are taken up and put to work in practical experiments in my work, as my co-authors and I look to art, design, architecture, cartography, cinema, etc., where we find examples of experimental practice. Further pursuing this theme, I draw from a three-day conference I cohosted in 2023 at MIXI, the Manhattan Institute for studies of STEM and the Imagination, entitled *Learning under Algorithmic Conditions*. The term *algorithmic condition* was used to reference the fundamental role of machine learning in shaping current images of learning, reason, explanation, pedagogy, cognition, doubt, expertise, and judgment. The reference to our algorithmic conditions also references our increasing reliance on machine learning methods in social sciences and underscores the indeterminacy and speculative nature of our investment in computational practices wholesale. I am feeling that urgency around our algorithmic condition now more than ever, and have tried to include chapters that deal directly with it. My aim is to stay focused on the play of contingency at the heart of speculative-mathematical models of chance – in other words, to separate statistics from mathematics, so we can discern the distinctive manner by which chance is formalized in current platforms and statistical methods, and thereby find ways of resisting and repurposing new media. We need to look under the hood of our technology, to deconstruct it's algorithmic-material structure, trace its network of reliance, and come to understand it. Exposing how something works is a kind of resistance. But I am all too conscious of the practices of control in digital governance and bodily subjection in digital life, and the narrow bandwidth for such resistance.

I was a fiction writer prior to entering academia and wrote extensively in the 2000s about the possible use of speculative and creative methods in the social sciences. More recently, and as part of my work in the Manifold Lab, I published a short science fiction story depicting a laboratory where human "qualiants" entered and were reassembled into an organic more-than-human mixture (de Freitas, 2017). The story was intended as a humorous depiction of a biosocial lab gone bad (or good?) because of an unanticipated accident in the material infrastructure of the "smart" laboratory, due to flooding from a heavy rainstorm (all too typical in Manchester!). The laboratory was laced with machine learning software, as a way of animating the learning environment. Following David Roden (2014), I wrote the SF story as a speculative thought experiment, and as a way of exploring the weirdness of posthuman methodologies. Titled "Nonhuman findings from the laboratory of speculative sociology," the story refused to start with any image of human sociality, and used the tactics of SF to explore the limits of inquiry. After the paper won an award at the *European Congress of Qualitative Inquiry*, a colleague came forward and laughingly said that she was disappointed to learn that the laboratory did not exist. But I replied: "Yes, it does! If not exactly as imagined." Indeed, this collection of chapters presents a series of theoretical forays and empirical experiments meant to continue the work of that SF story. In addition to the detailed theoretical rationale for such work, this book offers examples of projects that are (a) politically and materially engaged with digital transformations of life and culture, (b) attentive to the atmospheric and distributed eco-sensory nature of experience, and (c) explicitly focused on the role of contingency and speculation in experimental modes of empirical inquiry.

Computational cultures

This book is held together by a sustained scrutiny of practices performed in the name of knowledge production, seeking methods that might be well suited to algorithmic and Anthropocene conditions. My hope is that these essays offer concrete examples of a posthuman social science practice along with detailed theoretical arguments in support of such work, where the power of speculative thought and experimentation becomes an essential ingredient of inquiry. Inspired by STS thinkers like Latour (2015) and Barad (2007), I am critical of dominant social research methodologies that assume a particular concept of the social in advance and refuse to recognize the radical ingenuity of life, the potentiality of matter, and the creative force of activity. This collection of essays shares a commitment to speculative experiments that might help us pursue a social science of more-than-human associations, allowing the very concept of the social to mutate and emerge as radically new. My interest is in breaking with a social science that comes ready-handed with coding devices that cannot reckon with the creative and dangerous forces of worlding. The overarching aim is to draw attention to new qual-quant mixtures of meaning in our midst, to open the black box of software, and to argue that contingency is inherent to all methods. I believe this allows for a social science that might reclaim and rewrite the meaning of digital data, wrestling it away from surveillance capitalists. This book argues that critical and creative engagements with digital technology and generative algorithms have become urgent as data is increasingly "mined" by governments and corporations.

Mike Savage and Roger Burrows published "The coming crisis of empirical sociology" in 2007, pointing to the shifting ground of social science since the 1990s, due primarily to changes in technology. Directing attention to the "social life" of research methods, they describe an important methodological shift or crisis in conventional empirical methods. They recount how research methods once dominant in sociology – the sample survey, the in-depth interview, and ethnographic observation – are no longer well suited to current computational culture and the proliferation of both social transactional data and micro-bio data. Savage and Burrows (2007) claim that these methods are now seen by many as obsolete, in the midst of our current deluge of digital "social" data. Such data is automatically uploaded when we use our phones, computers, banks, coffee shops, toilets, etc.; then packaged, processed, and disseminated across widespread circuits of connectivity. Digital data of this kind has been labeled "passive data" because it is all too often "the digital by-product of the routine operations of a large capitalist institution" (Savage & Burrows, 2007, p. 887). The data deluge is thus "reassembling social science methods" and forcing social science to rethink its role in the study of the social, as digital devices are "increasingly the very stuff of social life" (Ruppert et al., 2013, p. 24):

> If digital devices mediate and are in considerable measure the stuff of social, cultural, economic and governmental lives in contemporary northern societies, then what does this mean for our methods for knowing those lives?
>
> *(Ruppert et al., 2013, p. 24)*

Recent generative AI algorithms are widespread in research today, furthering the digital impact on social inquiry. These algorithmic models operate – much like other devices in the history of empiricism – to actualize relations and connections that are otherwise beyond perception. Research is always about making visible and manifest new relations and new agencies that were not previously known. The artisan might work with the singularities of a particular slab of marble, while the digital artisan will need to find a way to work within the fissures and veins of databanks, following the swirls and streaks across the stockpiles of human traces. Digital life changes the way of the artisan, as such data is a new kind of materialization of a "traceable social" rendered newly visible, and thus actively involved in the production of a new imaginary and ontology of the social (Latour, 2007).

The current concern with the limitations of conventional social research methods is linked to what Braidotti (2013, 2022) calls our "posthuman" historical moment, and the realization that researchers must find better ways of studying a more-than-human society. Research methods that continue to center human intentionality and reinstate human exceptionalism tend to keep us stuck in bad methods, despite the fact that scholars are increasingly situating their work within the Anthropocene or what Haraway calls the "Capitalocene" or "Chthulucene" (a geo-political rearrangement of life, earth and power). For Haraway (2016), the challenge is to study complex systems and ecologies of material-cultural practices, to study distributed mindedness or sympathy in creative processes of worlding. For Colebrook (2011, 2014), the challenge is to do this without romanticizing the ground or environment in which new methods emerge (without reinserting the human as the ultimate sole benefactor). Perhaps this new kind of inquiry will in the future have to go by some other name – perhaps it will be a new kind of climatology or immunology of natureculture. Notably, the quantum figures prominently as a new paradigm for the social sciences, as various thinkers turn to contemporary quantum science to rethink future methods (or non-methods) that buck the Newtonian deterministic models that shaped our early social sciences (see for instance, Coleman & Ringrose, 2013; Colman & Van der Tuin, 2024; Mazzei & Jackson, 2024; Wendt, 2015). The quantum paradigm might open onto different ways of doing neural network architecture and reshape our understanding of cognition (de Freitas, 2025). Recently formulated methodologies – such as software studies, investigative aesthetics, and critical cartography – have emerged in the last decade, to draw on open-source data but also to remap the datafied earth. It seems highly likely that social science will become unrecognizable to itself in these shifts, and perhaps my efforts here will have indeed tracked the social life (and death) of its emergent empirical methods.

In 2008, the chief editor of the popular culture magazine *Wired* infamously declared that "big data makes the scientific method obsolete." This provocative comment concerns the monumental shift across the sciences as they become increasingly computational and data-driven. Commenting on computational biological methods, Stevens (2013) similarly states that the common Baconian image of the scientific method as hypothesis-driven is "giving way to hypothesis-free experiments and data collection" (p. 66). The data deluge and information explosion are so extensive, we have become enamored of AI methods that are insanely resource-heavy and wasteful of energy. These data-driven methods mark the dominance of neural network algorithms that mobilize naïve Bayesian probabilities (weights) to incrementally fudge their way to a verified hypothesis – we might characterize this as a kind of obscene incrementalism and reverse probability method (Joque, 2022). The dominance of these artificial neural network approaches to AI since the 2010s has reinforced the idea that theoretical constructs are emergent and fundamentally correlative (rather than causal), so that all structure is said to be inferred from patterns in data. This desire to be free of the responsibility of theory is a behaviorist's fantasy, and a delusional fantasy that the truth lies out there in data and it's only a matter of discovering it. My aim in this book is to adumbrate methods more suitable to a posthuman social science, methods that might better attend to the digital datafication of life without the reification of statistical science. We need to be better equipped to address the dangers of an "ecological rationality" that naively equates computation with primordial indeterminacy. One thing this book aims to achieve is to expose the specific power of the hypothetical and the force of contingency in research methods. As Parisi (2017) warns, big data methods underscore the links between affect and reason, whereby a "chain of contingencies becomes the driving force for decision-making actions." We can't deny the link between affect and reason, nor deny the fundamental role of computation and the quantitative in material relationality, but we need to resist crude materialisms that affirm the coincidence of matter and computation, we need to keep our eye on the incomputable.

The incomputable

How might we then develop methods that grapple with these limits, methods that diverge from convention and respect the incalculable? Patti Lather seeks a "multiply-sited subject" characterized by "heterogeneity, irreducible particularities, and *incalculable* differences" (Lather, 2017, p. 103, my italics). In discussing the work of Elizabeth Povinelli, Lather lauds her attempts to mix methodologies, combining GPS and geocoding with totemic Indigenous land praxis, creating a mixed methods approach that includes a cosmological gaze and a critical cartography. She describes this work as "in excess of intersectionality" in its "attention to multi-directionalities, post-human bodies, intra-actional networks, contingency, non-mastery, and *incalculables*" (Lather, 2017, p. 326, my italics). Lather also evokes the incalculable as part of the promise of new materialist methodologies that enact "post-post validity," a validity that would be "inexact, and yet rigorous, *incalculable*, flowing along the lines of Deleuzian transcendental empiricism" (Lather, 2017, p. 12, my italics). This future methodology inherits from traditions and works their ruins, but also invents the new through mixture and hybridity. The incalculable seems to ground this proposal: "we inherit and invent, each time anew, toward the something to come that is already at work, *incalculable*" (Lather, 2017, p. 311, my italics). Lather, who was concerned with a delusional science that saw itself in terms of "protocols of calculation" (Lather, 2017, p. 289), emphasizes the "volatility beyond calculation" (Lather, 2017, p. 323).

I am interested in how we might put flesh on her endorsement of the incalculable today, as we pursue new methods of inquiry that take up the digital. After recent work in software studies, how might we respond to her query: "What would the "incalculable subject" look like as a counter to neoliberal and Big Data efforts to count and parse, capture, and model our every move, a subject outside the parameters of the algorithms?" (Lather, 2017, p. 345). Lather's trenchant critique of experimental or quasi-experimental protocols and random-assignment experiments in the social sciences draws attention to the ways in which chance – in all its unruly contingency – is tamed and quantified in statistical knowledge production. It's important to emphasize how statistical thinking has come to further dominate computing practices. I draw attention here to how Lather's critique targeted the 20th century harnessing of the *random* (as the statistical naming of chance), because I want to emphasize how 21st century research methods continue to leverage the random, but in new ways. By staying tuned to how chance, contingency, and randomness are clothed in new ways, we can begin to envision new manners and styles for pursuing Lather's "getting lost" approach to the incalculable. My book continues Lather's attention to the incalculable, and links it to current computational paradigms. I am thus shifting the register slightly, and translating her "incalculable" to *incomputable* as I take up and grapple with the limits of software methodologies. In doing so, I hope to carry on her project of thinking about the quantitative as a moving target. She realizes all too well that the quantitative is something that is imbricated with the qualitative, and thus no more or less ontological. In other words, if there is a line that separates the computable and the incomputable, it is a moving line, an undulating fold in the fabric of life, be it digital or organic.

I learn from Lather that the future of qualitative and quantitative modes of engagement are themselves unscripted, as the two are reassembled in different media over time. In other words, the conjoined in/calculable is itself an indeterminate and opaque mixture, surfacing differently depending on our material conditions and mathematical practices. Today's digital media have to be examined for how they newly achieve that reassembling, and how those fighting for political liberty and equitable inclusion might reassemble this qual/quant differently. The history of technology and media shows how delineations regarding the border between the computable and the incomputable reshape the territory of life, labor, and learning. As Galloway (2021) suggests, computation is a historical material practice, and it can thrive or atrophy; domains come in and out of the realm of the computable. How should we manage this forever-changing presence in our lives? The last century has witnessed a global colonization through *digital* computing, as Turing's "universal machine" sees computation everywhere in the form of discrete-state machines. And yet there is always some sort of lurking dark matter, where that which is incomputable or incalculable, refuses to succumb. What exactly is the incomputable today? How is it mixed up with the computable? Following Galloway (2021), I will put forth here three different claims about our current configuration of the incalculable or incomputable.[1] These three claims are that the incomputable is found in (a) analog life, (b) rational paradox, and (c) practical material limits to calculation. All three of these answers offer some insight, as we try and navigate our technical milieu and the specific nature of contemporary computational cultures.

The first answer correlates the incomputable with the concept of "analog" which refers to continuous media rather than discrete and digital media. The analog is said to be closer to the body and organic life; it designates the flesh, the causal instinct, the intuition, affect, and aesthetic of "lived experience." Accordingly, *the body* as analog being is ultimately incomputable, in all its unruly and unmastered contingency. This first answer helps us contrast the continuous flow of organic life to the pulsing flux of the digital, or contrast the continuous variance of intelligence with the binary truth value of discrete state machines. But does this list of *flesh, affect,* and *aesthetic experience* capture the incomputable? Not quite. We need to ask ourselves what is lost when we install a too simple binary between the fleshy body and the digital machine, or between the continuous and the discrete. Haraway's cyborg manifesto was meant to help us overcome this too simple binary. How might computability be something that is misrecognized in these oppositions? How might these distinctions actually hinder our ability to imagine the incalculable subject that Lather seeks?

Yes, computing machines are alphabetic, in that they are all about reading, writing, and symbolic code (Sack, 2019), but they are also counting machines. Discrete state machines are logic machines, performing acts of sorting and branching. But they are deployed as counting machines, and counting allows them to produce hypotheses. What exactly is counting? What philosophically can we say about counting? It is a kind of calculating that seems cross-cultural and also more-than-human. Counting is a kind of calculating that is not in itself bad or destructive. If we reflect on the nature of counting and look at how children learn to count, we can see that counting is a kind of repeated ritual, an act of "forging a unit, then iterating the unit in series" (Galloway, 2021, p. 20). For instance, when weaving on a loom or programming a computer, or other situations where the conditions are ripe for such repetitious labor, we repeat this gesture of counting, calling on number as though it were a bottomless pit or a magic porridge pot. The work of calculating is thus achieved through an iterative *multiplicity* of repeated actions, a cyclic ritual of rendering *the next number*. In counting, even if the count is over, there is always a next, always another future or potential number. This kind of *ordinal* counting is not the cardinal counting of pinning down the size of a set. Ordinal number, as Deleuze and Guattari state (1987), is an ambient circulating number, moving like wild-fire rather than covering a space like a blanket. This ritual act of indefinite counting forges an enfolding unfinished multiplicity, performing the repetition that engenders the very idea of *the next*. Counting throws number into the fire and the sparks rise up. I am emphasizing here that number has profound ontological dimensions, and that the act of counting is a strange kind of repetition that is more-than-human. However, I am not ignoring the fact that calculation is an instrumental mode of governance and control, and that counting is also a fundamental ritualistic act of coercion in cultural and material collectivity (de Freitas & Sinclair, 2022). I want us to appreciate how number is both these things, that number works through different kinds of power. This claim is not meant to undermine the political critique of computational cultures, but simply to affirm the equally important insight that numeracy and counting are important political acts that bring forth cultural and political multiplicity – as Hardt and Negri (2004) suggest, the socio-political multiplicity is always by necessity *numerous*. In number there is multiplicity. This is part of its power. I believe that our very understanding of the political, and even the possibility of the political, depends on a concept of number, and that we can't simply blame number for all our bad computational practices. Galloway (2021) cites Euclid's notion of arithmetic numbers as "plethos" (plethora) (πληθσζ), which is often translated as "multiple" (p. 21).

The second answer to the question as to what is the incomputable refers to "rational paradox" and the limits of what I will call *positivism*.[2] Rational paradox haunts the positivist dream of a logical and axiomatic foundation on which science should be based, as articulated by mathematicians and logicians like Gottlob Frege, Bertrand Russell, and David Hilbert in their work on mathematical logic and proof theory. The limits to such foundational projects were uncovered by various logicians and mathematicians in the early part of the 20th century, when the epistemic foundations of mathematics and science were shown to be inherently unstable and haunted by paradox. Kurt Gödel's 1931 incompleteness theorems demolished the dream, demonstrating how truth and provability were not always coupled, even in the simplest of arithmetic worlds (Kennedy, 2015). He showed the existence of the incalculable in mathematical formalisms, undermining the aim to offer absolute foundations for quantitative truth. This destabilizing of epistemic certainty pointed to the existence of truths that elided the rules of inference. Gödel showed that there would always be truths we could not prove and that no consistent axiomatic system can prove its own consistency. He destabilized the dreams of logicism and the claim that mathematics was secondary to logic. Galloway (2021) uses the term *rational paradox* to refer to this inherent paradoxical failure of pure reason to furnish epistemic certainty. It underscores the kind of fallibilism at the heart of dreams of axiomatic closure.

The idea of rational paradox was already well developed in the 1970s by various philosophers. Poststructuralist theorists like Jean Lyotard (1979) took up Gödel's insight in *the Postmodern condition: A report on knowledge*, to critique representation and correspondence theories of truth in the social sciences. In like manner, more recent theorists of the incomputable, like Beatrice Fazi, cite Gödel's incompleteness theorems along with Alan Turing's halting dilemma, to argue that the incomputable is always lurking in any attempt to calculate anything. This affirms the persistence of incalculables. One can extend the claim that inherent incalculables haunt mathematics and find numerous historical examples; irrational numbers were considered incalculable by ancient Greek geometers, for whom the world was assumed to be composed of commensurable proportions ("rational numbers"). Even then, in ancient Greece, the realm of the incalculable was dangerous – Hippasus was allegedly drowned for pointing out how the diagonal of a square could not be expressed as the ratio of two integers. And notably, the ancient Greek solution to their dilemma was a *differential theory of averages* that iteratively approximated the unruly irrational number and gave them something they could work with, much like our current investment in neural network models of incremental computation. But it's important to know that Gödel's proof depends entirely on number theory in order to prove the weakness of logic! Number itself, for Gödel, was not the target of his destabilizing proof (Goldstein, 2005; Nagel & Newman, 2001).

The third answer to the question of what is the incalculable points to the actual pragmatic limits of calculation and computation. This is because acts of counting and measuring are literal material practices that consume time and energy. Computing consumes energy, and calculating is labor. There are material limits to what can be calculated in any given time or place, as resources frame the conditions of the computable. Quite literally, the incomputable lies beyond those material limits. With relevance to our current context, note that large language models (LLMs) and various other kinds of neural network computational models require excessive amounts of energy to make even the most mundane inference. They are resource consumers, using billions of parameters and massive amounts of data, and thus their computing practice is problematic in part because they consume obscene amounts of lithium, rare earth minerals, water, and all kinds of other physical resources. The incomputable in this case points to the physical limits of our computing power, and thus emphasizes the way that computation is a physical process, and that software runs on hardware. Our best examples of AI today are considered wasteful and highly inefficient machines precisely because they require so many resources (Chollet, 2019). Children learn how to generalize from only a few examples, and they do so with more agility than current machine learning models, and are thus more efficient learners. Questions of resource depletion and the environmental cost of various computing practices have to become much more central in our debates about the limits of calculation and the incomputable (Crawford, 2022).

These three answers to the question of the incalculable are helpful, if insufficient. I would like to add a fourth that pertains more explicitly to the role of chance. Lather's critique in the early 2000s of random trial research and sloppy applications of statistics must be reframed today, as we witness the methodological absorption of a new kind of randomness. Her critique of classical parametric statistical models used in the social sciences hit the mark, but today algorithmic reason deploys the random variable in new ways. We have moved beyond Lather's original target of random control experiments and are working with an entirely different way of mobilizing the random. Indeed, randomness is now mobilized in ways that are almost diametrical to how they were mobilized in earlier kinds of statistical research in the social sciences. Lather herself asked her audience time and again to attend carefully to the new algorithmic conditions that were emerging (de Freitas et al., 2016). The previous approaches to mathematizing chance treated the random as that which was "outside noise" or perturbations that troubled the model. Past scientific paradigms saw "the control" population as the calculable, and the random noise as that which muddied the clarity of inquiry. Our new algorithmic condition, however, has internalized such noise, so that we see a new kind of operational randomness built into the models. Chance is no longer treated as that which reduces the certainty of knowledge but is instead packaged as contingency and entropy, which becomes the engine of inference (Parisi, 2017). Practices that are internal to algorithmic work – random vectors, noise modulation, error redistribution, generative hallucinations, etc. – have all become part of what we can call *procedural randomness*. As Galloway (2021) claims, "in a sense, randomness and contingency have become fully industrial" (p. 20). In terms of methodological futures and the pursuit of what Lather called the incalculable subject, I agree with Galloway that there is no choice here between critique and computation. We must do both. But as we glimpse a future of quantum computing and quantum social science, there are new inflections of chance on the horizon (Wendt, 2015). As I discuss in Chapter 14, the force of generative AI technologies, increasingly dominant in our computational culture, needs careful philosophical and material scrutiny.

The diluvian metaphors of flood and deluge abound in our current computational culture, as we flow with the virtually infinite connectivity of the internet, drown in the excessive number of choices for every decision, surf the tidal sea of an inhuman number of selfies, or track viral fake news circulating in a near-infinite swamp of misinformation. My use of the word "virtually infinite" or "near infinite" is important, for it often seems that the metaphors of deluge and flood are used to conflate the discrete and finite world of digital data with the continuous flow of the infinite. And I appreciate Sarah Pourciau's (2022) suspicion about this digital ocean, and the way in which it shuns the actual infinite. She emphasizes how the turn to digital technology and discrete-state machines, as formulated in the work of Alan Turing, is a risky and negligent "shrug" that allows every life to become computable, "every kind of numerical reality that we might ever want to describe, from infinitesimal calculus to Fourier transformations and beyond, all without requiring any illicit recursions to the (much "bigger") resources of the continuum" (p. 255). This is the kind of numeracy that Turing ushered in. But it's equally important to recall different historical threads of *minor* mathematics, which always grapple with the aporia between the discrete and the continuous, where the incomputable often erupts. This is why we need to interrogate the actual statistical reasoning that has been packaged in our current machine learning AI methods and expose its tenuous grasp of the real. Inasmuch as the digital ocean is different from other oceans, it is because chance is now packaged in ways it wasn't in the 19th century.

Chapter synopses

Chapter 2 focuses closely on the theoretical foundations of posthuman science, offering a detailed analysis of Bruno Latour's concept of the *metamorphic zone* of terrestrial life, within the current climate crisis and environmental "cosmocolossus." I explore his proposal for a risky diplomacy in the Anthropocene, as a way of dealing with the increasingly tense relations between polarized and weaponized perspectives about truth (I seek the truth of the relative, rather than the relativity of truth). I show how his work is extremely germane to our current conditions, as he envisions a form of scientific practice that involves the invention of equipment (apparatus) that makes perceptible (principally to humans) the existence of nonhuman agencies, thereby expanding the opportunities for alliances and a pluralist ecology, and ultimately assembling another more-than-human political body. I also discuss critiques and concerns regarding the specifics of this proposal. In this chapter, I explore (a) the use of the term "climatic regime" and the complex ways in which social science must think within the Anthropocene, (b) Latour's emphasis on agency, animacy, and immanence as part of his "practical metaphysics," (c) his proposal for a risky terrestrial diplomacy and compositional peace under post-truth conditions, and (d) the way in which the "end times" is conceived as part of a new historicity and spatial imaginary. These topics together link his ontological commitments with his interest in the politics of empiricism and knowledge-pursuing practices. Latour proposes a kind of posthuman diplomacy for the "end times" as a way of dealing with our "post-truth" situation, and the increasingly tense relations between various polarized and weaponized perspectives. Such an endeavor involves a remixing of politics and earth, reckoning with the specific historical challenges we face today. In this essay, I dig into the ontological commitments underpinning Latour's image of science, and I explore the political implications of his proposal for an Anthropocene diplomacy that embraces "a Politics in which the role of Nature is explicit" (Latour, 2017, p. 47).

Chapter 3 investigates the prescient work of posthuman sociologist Gabrielle Tarde (1843–1904), and his relevance to our current use of big data methods in a "sociology of association" he called monadology. Both Gilles Deleuze and Bruno Latour ascribe to variants of monadology in their proposals for a new empiricism, drawing extensively on Tarde's audacious posthumanist 1895 text *Sociology and Monadology*. In this chapter, I discuss how monadology helps us rethink research methods in these digital times. I argue that a *fractal monadology* re-assembles the fold with the digital, the continuous with the discrete, and ultimately offers a philosophical foundation for contemporary social science research. This chapter argues for a new kind of qual-quant methodological mixture suitable to current computational cultures and algorithmic conditions. The classical methodologies of anthropology and sociology have had their conventions troubled, problematizing assumptions regarding the *unit* of analysis and the extent to which data are owned or even correlated to a definitive organic and individuated subject customarily referred to as "human." These methodological shifts demand a more careful consideration of the historical lineage of empiricism and its relation to the history of science more generally.

Chapter 4 opens up discussion about another key thinker from Science and Technology Studies, Karen Barad, whose writing on quantum ontology has had a huge influence on the shaping of new materialisms and posthumanisms. In this chapter, I focus on her attempts to reformulate the concept of relationality. The aim is to show how Barad's work articulates a new kind of empiricism for the social sciences by reclaiming the creative and speculative force of *experimental practice* and by re-centering the philosophical problem as a source of inquiry. Relationality is redefined through discussions of diffractive apparatus, more-than-human performativity, and the "polymorphous perversity" of the matter-meaning mixture. This chapter also considers the criticisms of her characterization of a new quantum empiricism, found in the work of philosophers Graham Harman and Slavoz Zižek. I interrogate the term "relationality" as it has come to saturate the discourse of cultural studies. Notably, Barad does not adequately address the absolutely central role of probabilistic reasoning underlying her more qualitative description of queer quantum behavior. This silence needs attention, as it seems likely that some notion of generative chance, as marshalled within probabilistic forms of activity, may be the key to this new kind of relationality, if not to our current computational cultures, as discussed in Chapter 14.

Chapter 5 shows how conventions of contemporary ethnographic video data are linked to the history of scientific cinema and the tracking of embodied movement. This chapter critiques the use of video analysis software that automatically sorts and codes video data, while offering up a different and much more artful way of deploying digital tools. I use the ideas of Gilles Deleuze – in particular his writing on cinema and the moving image – to explore an early generative algorithm that disassembled digital images, which might have otherwise been interpreted according to standard phenomenological theories. One of the key claims of Deleuze is that the "time image" is a kind of crystalline contraction and expansion of intensity, where the human body becomes a "shock of forces" rather than an integrated and self-knowing subject. I argue in this chapter that such an analytical approach to video data taps into new generative paradigms in digital technology and opens up new ways of studying events. I've continued this focus on video research in another recent chapter, co-authored with my colleagues, called *Bending time through participatory video methods: Reassembling urban kino-cinema with young people* (see Hickey-Moody et al., 2025).

Chapter 6 stakes out the political risks of a new biosocial paradigm that engages with digital sensor technology in social research and brings together important theory developed in media studies and feminist philosophy – in particular the work of Mark Hansen in media studies and the work of Elizabeth Wilson (2015) in feminist theory. This chapter also works closely with the insights of John Protevi, who is a Deleuzian scholar and has written about developments in embodied cognition, mapping the links between Deleuze and the 4E movement (embodied, embedded, enacted, extended). This paper presents a theoretical framework for rethinking biosocial research and critiques examples of research methods used in perception studies and computational affect studies. My approach moves away from theories of agent-centered perceptual synthesis (on behalf of a perceiving organism) and toward a more expansive understanding of the biosocial learning environment. The focus is on sensor technologies that track sensation below the bandwidth of human consciousness. Following Wilson, I show how the sympathetic nervous system is involved in *intense* eco-social contractions that take place above and below the thresholds of representational thought and individuated emotion states. I argue that there is an urgent need to reclaim this kind of biodata as part of an unequally distributed worldly sensibility and to thereby undermine more narrow reductive readings of such data. The paper explores the biopolitical implications (Rose, 2013) of recasting biodata in terms of trans-individual inhuman forces, while continuing to track the distinctive power of humans.

Chapter 7 continues to explore the potential use of digital technologies in ways that push back at the mainstream applications, with "surrealist" experiments. Written with

Maggie Maclure and David Rousell, this chapter explores how digital technologies can be used to raise awareness about *surrealist* methods within the field of ethnography. We argue that surrealist experiments with technologies push "sensory ethnography" – which has become a dominant trope and methodology across the social sciences, largely inspired by the work of Sarah Pink – into a more posthuman and ecological approach. This chapter counters the humanist assumptions about sensation and sensory data, which continue to shape much of the work done in the name of sensory ethnography. Critically responding to the contemporary capture and redistribution of experience through digital technologies, we present a series of creative ethnographic experiments, framing them in terms of past surrealist experiments in ethnography. We draw on theories of the haptic eye and a/signification from Deleuze and Guattari to problematize conventional descriptions of sensation, in search of the intensive, the imperceptible, and the eco-sensory. Our focus in this chapter is on digital technologies as potential agents of surrealist ethnographic experimentation, with particular attention to how such surrealism can be part of a post-colonial research practice. We engage with digital devices as ecological agents, tracking intensive flows, layering affect, invaginating power, ramifying insight, and entangling concepts through playful inversion, pursuing a kind of surreal participation that avoids de-scription and conventional causal exegesis. The experiments derive their post-colonial power from the work of South African artist William Kentridge, with whom the work is in direct conversation.

Chapter 8 presents spatial research methods and software applications used to study atmospheric and affective data. The research methods in this chapter were developed as part of a project funded by the Economic and Social Research Council in the UK and extended into a sister project in the US, funded by the Spencer Foundation. In this chapter, Laura Trafi-Prats and I argue for a "software studies" approach whereby digital methods serve to shake up the apparent fixity of spatial rules, defamiliarizing and indeed alienating us from what feels given – essentially repurposing digital tools that are all too often simply reinforcing hierarchical and rigid oppressions. In our focus on school buildings, this chapter builds on and extends my earlier work on counter-movement in relation to the built environment (de Freitas, 2011). This chapter emerged in conversation with Lisa A. Mazzei and Alecia Y. Jackson, whose sage editorial guidance helped shape the ideas. Laura and I respond to the postfoundational condition of qualitative research by taking a cue from software studies and explore the *glitch fragility* of digital data. We suggest that social science research might become a kind of *software art* that reckons with algorithmic contingency, not only as lack or deficit in representation, but as an opening onto the trans-individual and pre-personal atmosphere of the built environment. We discuss a research project about the atmospheric conditions of school buildings, and a series of experiments that engaged with school building envelopes. Our investigative approach was collaborative, working with participants to reanimate the school milieu and reveal the incomputable dimension of lived architecture. We discuss how the *evental* nature of the building was lured out of stasis by the student engagement with various technical and sensory devices. This chapter responds to the editors' orienting questions regarding research that engages the political through attention to the contingency in research methods, while exploring the pre-personal and pre-individual.

Chapter 9 continues to share experiments from the ESRC-funded project and offers more innovative methods for studying lived architecture and learning environments.

Laura Trafi-Prats and I present a series of mapping methodologies that incorporate and apply Latour's ideas about metamorphic zones and risky terrestrial diplomacy. Building on the Latourian ideas explained in earlier chapters, our approach works directly with Latour's concept of the *Terrestrial* as a metamorphic zone that includes the built environment, with connections to contemporary architecture and design practices. We use techniques in speculative cartography and cosmography, creating maps and models that show how students negotiate the affective atmosphere and social-material ecology associated with school buildings. These methodologies were inspired by the work of Aït-Touati et al. (2022), in their book *Terraforma: A book of speculative maps*, for which Latour wrote the preface. We saw in this work a visual mapping approach for studying the emotionally charged environment of school buildings, with particular attention to situated movements and modes of engagement after the Covid pandemic. Drawing on digital-sensory methods and software arts, we present a series of speculative architectural models and maps.

Chapter 10 analyzes the power of speculative thought in new and old empiricisms. Co-authored with Sarah Truman, this chapter emerged during her postdoc at Manchester Metropolitan University. Sarah's research was focused on how SF might help displace some of the whiteness in school literature curriculum, while I was interested in how SF might help us explore science distrust, especially after 2016 and during the pandemic. We developed a collaborative way of reading and analyzing science fiction as cultural expression. In this chapter, we work with an SF novel called *Annihilation* by Jeff Vandermeer (2014), exploring what a new posthuman science might look like, if it were to be feminist and anti-racist, and emerge in the Anthropocene. We discuss the power of speculative fiction and try to answer the following two questions: (1) Are there scientific imaginaries evoked in speculative fiction that might be helpful in rethinking the nature of empirical inquiry? (2) In what ways does speculative fiction present alternative research methods for studying pluralist posthuman ecologies and new forms of worldly belonging? We frame our chapter as a response to the search for new empiricisms and transdisciplinary methods that have led many social inquirers to engage with 20th-century post-Newtonian physical science. This chapter supports efforts to rethink the social sciences beyond Newtonian paradigms of causality, responding to new materialist concerns that past theories of sociality have dismissed the vibrancy and animacy of the nonhuman material world (Dolphijn & Van der Tuin, 2013). This chapter explores the power of speculative fiction to help us rethink empiricism in posthuman ecologies of the Anthropocene, in the midst of post-truth conditions and growing science denialism. We foreground speculative fiction as a way to open up scientific imaginaries, rethinking the relationship between nature, technics, and human "sense" making. We show how such texts offer alternative images of research methods for studying pluralist ecologies and new forms of worldly belonging.

Chapter 11 continues the exploration of the power of speculative thought in science, but tackles the very difficult issue of trust in science, and returns to discuss the limitations of an onto-epistemology that is grounded in the concept of relationality. This chapter looks more directly at why people distrust science, and the ways in which that distrust has changed over time, as scientific practice has evolved and moved away from the 19th century concept of objectivity (Daston & Galison, 2010). This chapter also explores a set of dilemmas associated with relational ontologies in which contingency factors prominently, raising the methodological question: how should we come to know, and trust our knowledge of the world? I focus here on the role of mathematical abstraction in science, and the limits of empiricism under chaotic climatic regimes and argue that a pluralist "multiple earth" must include mathematical fabulation and other kinds of speculative abstraction. This draws on my work on recasting mathematical practices as material practices (de Freitas, 2021; de Freitas & Sinclair, 2022)

Chapter 12 argues that a particular image of the "infant mind" continues to shape epistemological assumptions in AI research, and that the term "computational thinking" has undergone fascinating transformations as it saturates education policy. Matt Curinga and I argue that the debate between empiricism and nativism can be traced to recent developments in AI, and that the emphasis on "computational thinking" must be interrogated and historically situated. Critics of deep learning AI frequently cite studies of child behavior and the capacity of infants to learn quickly after only one or few encounters ("one-shot agile learning") to support the claim that *a priori* cognitive symbolic structures are the foundation of human and machine intelligence. In this chapter, we show how the debate between neural and symbolic AI rests on a particular image of the infant mind, elaborated by Seymour Papert and Marvin Minsky at the MIT Artificial Intelligence Lab in the 1960s. We show how this image of the child's mind was displaced by the concept of "computational thinking" during the 2000s, when neural network AI began to flourish. Finally, we turn to current work in Software Studies, which we characterize in terms of a post-Kantian and posthuman epistemology, to explore how the "network" image has replaced all earlier ideas about knowledge, learning, and computing.

Chapter 13 presents the concept of *epistemic imaginary* as a way to pursue post-colonial methods of inquiry that break with destructive empiricisms. This chapter continues the focus on the power of speculative thought, by working with the ideas of the French Caribbean poet and place-philosopher Édouard Glissant (1997, 1999, 2019) as a guide in the murky poetic realm of *epistemic imaginaries*. Glissant shares with Deleuze key theoretical tropes, linking his archipelagic thought to Deleuzian ideas about space, power, and knowledge. I also discuss the theoretical turn to Glissant in the work of various contemporary thinkers – Achilles Mbembe, Kara Keeling, Zakkiya Iman Jackson, Kevin Quashie – who carefully supplement their post-colonial critiques of science and Enlightenment reason with an affirmation of the imagination and its role in epistemology. The term "epistemic imaginary" is used in this chapter to underscore the power of the imagination in sparking insight, shaping knowledge practices, and establishing epistemic authority. The *imagination* is the milieu within which reason habituates itself, fabricating hypotheses and anticipating empirical encounters. This complex polemic between the imagination and empirical "correction" gradually sustains habits of trust and epistemic certainty. An epistemic imaginary is thus a set of knowledge claims performed passionately as a truth warranting, a set of declarations that establishes positionality and situates perspective, perhaps substantiated by other thinkers to some degree, and often enshrined and then imposed on others beyond the initial pertinent boundaries. Working closely with the speculative and poetic image of thought developed in the work of Glissant, this chapter takes up timely concerns with post-truth conditions and the corporate cooption of the speculative imagination, exploring the dynamics by which an epistemic imaginary is always made manifest in a manifold of contested assertions, delusions, and speculative ventures, almost always situated in spatial politics. This chapter closes the book with a simple message made new with Glissant's unique Caribbean perspective on place: *Knowledge and truth-telling are habits born of the imagination in place.*

Chapter 14 looks directly at algorithmic Large language models (LLMs), and the ways in which scholarship is changing under these new computational models. The chapter explores the following questions: How are reading habits changing, as our technical milieu becomes increasingly digital and our reading increasingly automated? And what is the role of philosophy – and in reading philosophy – as we grapple with new technical modes of reading? Guided by philosopher Gilbert Simondon, this chapter compares the language heuristics of LLMs with human reading practices, revealing parallel and diverging technical tactics, with the aim of increasing our understanding of how and why these algorithms are part of our technical reality. This comparison moves beyond concerns with automation and alienation, using Simondon's notions of technicity and transindividuality to philosophically analyze the nature of collaborative reading in a distraction economy, and the extent to which transformer neural network models achieve an implicit embodied or grounded sense of language use.

The temporal fabric of research methods

Revolutionary change in science occurs when methodological rituals fall apart; we are witnessing the falling apart of both qualitative and quantitative methodologies in the social sciences, as our life-technic-world has once again metamorphized into a new computational culture. And as much as I am excited by the opportunities for a cosmic artisan, as Deleuze and Guattari (1987) suggested, I am all too aware that such efforts are both enabled and yet all too easily botched during such periods of radical transformation. The precarity in such conditions is pronounced – proceduralism is always waiting to take hold, with a tendency toward entrenchment and ossification. The fetish to measure and fix methods is always there. My effort in this book is to show how "measures" are themselves historical and open to change, that measures are relational, intensive, and metamorphic, and open to radical reconfiguration. If we can begin to grasp, as Barad suggests, that measures and numbers are fully materially implicated in a processual matter, we will become even more politically empowered:

> Theorizing, a form of experimenting, is about being in touch … The world theorizes as well as experiments with itself. Figuring, reconfiguring. Animate and (so-called) inanimate creatures do not merely embody mathematical theories; they do mathematics.
>
> *(Barad, 2012 p. 207)*

Granted, research methods often reinstate social orders and hierarchies, and heavily control, if not curtail, knowledge production; but if we want to reclaim digital methods for more creative efforts, so that computation does not just serve corporate interests, we must overthrow the reified procedures, while retaining this Baradian insight about the power of speculation and mathematical theories. Chapter 11 speaks directly to the many ways that mathematical abstraction plays a part in relational ontologies. We need to reclaim measurement and number as down-to-earth modes of *lived abstraction* (Massumi, 2014).

The question of temporality is crucial. Research methods are spatially and temporally situated, enfolding in time and place. The temporal dimension of our methodologies is often overlooked, despite its structuring and conditioning what we are able to say about our research. Key questions should always be asked: What is the proper temporal unit of analysis (a gesture, an encounter, a year, a life, a decade, a synapse)? What kind of presentism or futurity is entailed in descriptive and predictive research? How do the research tools actually curtail our ability to make sense of the duration of time? When do new research methods emerge in a culture and why? Manning (2016) points out that "In working as an apparatus of capture, method gives reason its place in the sun: it diagnoses, it situates, it organizes, and ultimately it surveys and judges" (p. 32). Research methods can thus be staid and oppressive, but they can also be innovative and artful, demanding *an attunement to the differential becoming of the event*. Open problems are invented through such emergent methods. Manning (2016) asks that we develop a new register to pursue this alternative way of doing research: "This new register is neither art per se nor philosophy: it is study, it is practice, it is speculation" (Manning, 2016, p. 13).

Looking at method through the lens of temporality allows us to identify core characteristics of inquiry more generally and, in so doing, highlight the ways in which "method" is a grounded and situated earthly practice (Manning, 2016). This is not to naturalize a method, but to try and understand how and why research methods emerge. This focus on the temporality of method is crucial as we consider our current data inundation and the supposed eclipse of theory. Following Serres (2017), we can state that an artisan practice becomes a method when it is framed by a larger temporal scale, separated off from the undifferentiated indeterminacy and heterogeneous rhythms of life writ large. Artisan practices become research methods through repetition, through the rhythm of their technicity, as the ideas-in-action are enacted and extended. The cosmic artisan reverberates and feeds forward into the differential system, structuring that which is near at hand with their inventive experiments. The Latin word *method* once simply meant path or "pursuit or following after" from "meta" (after) and "hodos" (system, way or manner of doing or saying) (Agamben, 2016). In this sense, method is a cultivated habit which is both active and passive, shaping life while also submitting to that which one investigates. Method is both *a going* and *a yielding*. Method always involves this heterogeneous flow. If researchers are to do more than impose parametric models on digital data, or tweak the weights of their neural networks, that is, if social science is to do philosophy, then the recording and analysis of data must become newly assembled, newly configured, according to newly queered relations. And this is why the social sciences must rethink the *temporality* of digital methods, to tap into different rhythms in their work (Mikulan & Sinclair, 2023).

The investment in a research method is often the wish that one could occupy the present as origin. A method presumes a point from which one sets out. This is the instrumental dream that we might "start" in a purified space, clear the slate and *commence* to make sense of something. In 1637, Descartes published his *Discourse on the Method of Rightly Conducting the Reason* with just such an agenda. Research method for Descartes channeled a problematic desire for an origin, to locate or begin the inquiry or intervention in some neutral "control case." This is the art of Cartesian method – one must begin, and such a beginning claims to first sweep aside the detritus of past methods. Many methods are built on such acts of erasure, acts that still and shush the rumbling of the event. In Descartes' case, determination of truth must first eliminate all sources of delusion to find the clear and distinct idea. His method of investigation exemplifies the methodical fantasy of an inventive discontinuity (to start from scratch).

Research methods are often aligned with colonialism, as a means of carving up space and time, reinforcing colonial land-grabs and empire-building. The Eurocentric whiteness and maleness of positivist science invested in methods that "start from scratch" as a way to appropriate and quell the power of blackness and otherness (Iman Jackson, 2020; Tuhiwai Smith, 2021). And the violent impact of such appropriation has continued in research methods that fail to respect radical difference, or pathologize it as less than human. As we discuss in chapters 7, 10, and 13, qualitative methods in the social sciences have historically invested in capturing (in the name of liberation) and misrepresenting indigenous and non-European peoples. A posthuman social science must carry that historical load. In rethinking method, we must destabilize the gender-conforming, heterosexual and racialized category of the human. Iman Jackson (2020) recounts the specific ways in which the black body has been animalized and othered from the human. Agamben (2004) reminds us that the "animal" has been denied within human "civilized" culture: "man is the animal that must recognize itself as human to be human" (2004, p. 26). In discussing postcolonial Caribbean homophobic and racist contexts, Gosine (2021) asserts that the animalization of Caribbean people has produced persistent racialized violence for 500 years. And yet he also says that a pivotal political move for a blackened body today is to *refuse* to respond to the racist call to prove oneself "human, not animal." In other words, refuse that question as ill-posed. He argues against a divide between human and animal that is "both cognizant of the history of racialized animalization … and an ecological commitment" (p. 8). The declaration "we are animal" in the midst of the current planetary crisis, he claims, is a timely political move that is needed, a way of reimagining the human, to push back at the heteronormative and racialized violence of colonial history. Chen (2012) similarly argues for an increased awareness of the animacies that constitute our embodied and emplaced trans-material being, while documenting the racial and queer mattering that fuels inequity. As discussed in Chapter 13, Jackson (2020) turns to poetic and artful interventions to reclaim the animal and the more-than-human. Keeling (2019) also exposes the anti-blackness of scientific inquiry, and yet seeks new speculative ways to engage with it: "Working with and through scientific and technical knowledge in the service of humanistic inquiry is a challenge of world historical significance" (p. 195). Warning of the emerging "necropolitics," Mbembe (2019) states:

> We are thus in the process of entering a new human condition. Humanity is in the process of leaving behind the grand divisions between the human, the animal, and the machine so typical of discourse on modernity and on humanism. Today's human is now firmly wedded to its animal and its machine, to a set of artificial brains, of linings and interfacings (*de doublures et de triplages*) that form the base of the extensive digitalization of its life.
>
> *(p. 179)*

The cosmic artisan must inherit this historical awareness and this more-than-human perspective, rather than declare itself the first. We need to trouble the image of method as a break with the continuity of time, and an attempt to restart time here and now. Temporality passes through the porous methods that claim to stop and start it, in the vain attempt to mark a "now" as the difference between before and after. In our search for innovative methods, we need to trace their historical lineage, to notice how they preserve a certain impurity of the previous methods, despite the apparent clear slate and origin they are meant to offer up (Stengers, 2018). Humbly, we need to expose the old methods subsumed by new ones – expose how the field of the learning sciences subsumes constructivisms, the field of post-qualitative methods subsumes ethnography, the field of machine learning subsumes classical statistics. New methods always recuperate and envelop previous methods, as much as they aim to hide that provenance in the "new." This is the method of method. A cosmic artisan must grapple with this strange process of repetition and difference. They must accept that the scientific method and knowledge production more generally *always entail this haunting remainder, this debt to time* (Serres, 2017).

Notes

1 Galloway (2021) uses the term "uncomputable" rather than incomputable. I have stuck with the latter, as it is nicely parallel in its construction to the term that Lather chose, the incalculable.
2 I have two reasons for preferring to use this term positivism to another that is often used – "Enlightenment reason." First, positivism is a name that was used to describe a particular theory of scientific knowledge production, which is recognizably absorbed into 20th century social sciences, unlike the vague notion of Enlightenment reason. Second, ample scholarship has shown that the Enlightenment is not a monolithic single perspective and includes complex counter currents, including the unruly baroque.

References

Agamben, G. (2016). *The use of bodies: Homo sacer IV, 2*. (Trans. Adam Kotsko). Redwood City, CA: Stanford University Press.
Agamben, G. (2004). *The open: Man and animal*. Trans. Kevin Attell. Stanford: Stanford University Press.
Aït-Touati, F., Arènes, A., & Grégoire, A. (2022). *Terra Forma: A book of speculative maps*. (Trans. A. DeMarco). Cambridge, MA: MIT Press.
Barad, K. (2007). *Meeting the universe halfway: Quantum physics and the entanglement of matter and meaning*. Durham, NC: Duke University Press.
Barad, K. (2012). On touching: the inhuman that therefore I am. *Differences: A Journal of Feminist Cultural Studies*, 23(3), 206–223. Revised and republished in Power of Material/ Politics of Materiality edited by Susanne Witzgall and Kirsten Stakemeier, 153–165.
Braidotti, R. (2013). *The posthuman*. New York: Polity Press.
Braidotti, R. (2021). *Posthuman feminism*. Hoboken, NJ: Wiley Press.
Braidotti, R., & Bignall, S. (2019). *Posthuman ecologies: Complexity and process after deleuze*. New York and London: Rowman & Littlefield.
Chen, M. Y. (2012). *Animacies: Biopolitics, racial mattering, and queer affect*. Durham, NC: Duke University Press.
Chen-Morris, R. (2022). Fantasy, scientific thought, and the end of baroque science. Online lecture. Available at https://buberfellows.huji.ac.il/event/fantasy-scientific-thought-and-end-baroque-science.
Chollet, F. (2019). *On the Measure of Intelligence*. arxiv.org. https://arxiv.org/abs/1911.01547.
Colebrook, C. (2011). *Essays on extinction*. Open Humanities Press. Creative Commons. Available at: www.openhumanitiespress.org/books/titles/sex-after-life
Colebrook, C. (2014). Socially invaded: The biosocial subject. *Environment and Planning: Society and Space*, 32, 556–570. Available at https://journals.sagepub.com/doi/pdf/10.1068/d3203rev (accessed 30 October 2017).

Coleman, R., & Ringrose, J. (2013). *Deleuze and research methodologies*. Edinburgh: Edinburgh University Press.
Colman, F., & Van der Tuin, I. (2024). *Methods and genealogies of new materialisms*. Edinburgh: Edinburgh University Press.
Crawford, K. (2022). *Atlas of AI: Power, politics, and the planetary costs of Artificial Intelligence*. New Haven, CT: Yale University Press.
Daston, L., & Galison, P. (2010). *Objectivity*. Princeton, NJ: Princeton University Press.
de Freitas, E. (2011). Parkour and the built environment: Spatial practices and the plasticity of school buildings. *Journal of Curriculum Theorizing*, 27 (3), 209–220.
de Freitas, E. (2017). Nonhuman findings from the laboratory of speculative sociology. *Minnesota Review*, 88, 116–126.
de Freitas, E. (2021). Mathematics in the middle: The relationship between measurement and metamorphic matter. *Matter: Journal of New Materialist Research*, 2(2), 1–24.
de Freitas, E. (2025). Quantum cognition and the limits of classical probability models. *Computational Culture: A journal of Software Studies*, 10: http://computationalculture.net/quantum-cognition/
de Freitas, E., Dixon-Román, E., & Lather, P. (Eds.) (2016). Alternative ontologies of number: Rethinking the quantitative in computational culture. *Cultural Studies-Critical Methodologies*, 16(5). https://doi.org/10.1177/153270861665575.
de Freitas, E., & Sinclair, N. (2022). Proportion, analogy, mixture: Unearthing mathematical measurement practices. In T. Ingold (Ed.), *Knowing from the inside: Cross-disciplinary experiments with matters of pedagogy* (pp. 63–80). London: Bloomsbury Publishing.
Deleuze, G. (1994). *Difference and repetition*. (Trans. P. Patton). New York: Columbia University Press.
Deleuze, G., & Guattari, F. (1987). *A thousand plateaus: Capitalism and schizophrenia*. (Trans. B. Massumi). Minneapolis, MN: University of Minneapolis Press.
Dolphijn, R., & Van der Tuin, I. (2013). *New materialism: Interviews and cartographies*. Open University Press. https://www.openhumanitiespress.org/books/titles/new-materialism/
Galloway, A. (2021). *The uncomputable: Play and politics in the long digital age*. New York: Verso Books.
Glissant, E. (1997). *Poetics of relation*. Ann Arbor, MI: University of Michigan Press.
Glissant, E. (1999). *Caribbean discourse: selected essays*. (M. Dash, Trans.). Virginia, US: University of Virginia Press.
Goldstein, R. (2005). *Incompleteness: The proof and paradox of Kurt Gödel*. New York: W.W. Norton & Company.
Gosine, A. (2021). *Nature's wild: Love, sex and law in the Caribbean*. Durham, NC: Duke University Press.
Hansen, M. (2015). *Feedforward: The future of twenty-first century media*. Chicago, IL: University of Chicago Press.
Haraway, D. J. (2016). *Staying with the trouble: Making kin in the Chthulucene*. Durham, NC: Duke University Press.
Hardt, M., & Negri, A. (2004). *Multiplicity: War and democracy in the age of empire*. London: Penguin Books.
Hickey-Moody, A., Pihkala, S., Coombs, G., & Willcox, M. (Eds.). (2025). *New materialist affirmations: Creative research interventions in methods and practice*. Edinburgh: Edinburgh University Press.
Hird, M., & Yusoff, K. (2019). Lines of shite: Microbial-mineral chatter in the anthropocene. In R. Braidotti and S. Bignalle (Eds.), *Posthuman ecologies: Complexity and process after deleuze* (pp. 265–82). New York: Rowman & Littlefield.
Jackson, Z. I. (2020). *Becoming human: Matter and meaning in an antiblack world*. New York: New York University.
Jagodinski, J. (2023). Virtuality. In N. Wallenhorst & C. Wulf (Eds.), *Handbook of the Anthropocene* (pp. 993–997). Berlin: Springer.

Joque, J. (2022). *Revolutionary mathematics: Artificial intelligence, statistics and the logic of capitalism*. New York: Verso Books.
Keeling, K. (2019). *Queer times, black futures*. New York: University of New York Press.
Kennedy, J. (2015). Kurt Gödel. *Stanford Encyclopedia of Philosophy*. https://plato.stanford.edu/entries/goedel/
Lather, P. (2017). *(Post)Critical methodologies: The science possible after the critiques*. New York: Routledge.
Latour, B. (2007). *Reassembling the social: An introduction of actor-network-theory*. Oxford: Oxford University Press.
Latour, B. (2015). *Reassembling the social: An introduction to actor-network-theory*. Oxford: Oxford University Press.
Latour, B. (2017). *Facing Gaia: Eight lectures on the new climatic regime*. Cambridge: Polity Press.
Latour, B. (2018). *Down to earth: Politics in the new climatic regime*. New York: Polity Press.
Lyotard, J. (1979/1984). *The postmodern condition: A report on knowledge*. (Trans. G. Bennington & B. Massumi). Minneapolis, MN: University of Minnesota Press.
Manning, E. (2016). *The minor gesture*. Durham, NC: Duke University Press.
Massumi, B. (2014). *The power at the end of the economy*. Durham, NC: Duke University Press.
Mazzei, L., & Jackson, A. (2024). *Postfoundational approaches to qualitative inquiry*. New York: Routledge.
Mbembe, A. (2019). *Necropolitics*. (Trans. S. Corcoran). Durham, NC: Duke University Press.
Mikulan, P., & Sinclair, N. (2023). *Time and education: Time pedagogy against oppression*. London: Bloomsbury Publishing.
Nagel, E., & Newman, J. R. (2001). *Gödel's proof*, 2nd edition. New York: New York University Press.
Parisi, L. (2017). Computational logic and ecological rationality. In E. Hörl (Ed.), *General ecology: The new ecological paradigm* (pp. 75–100). London: Bloomsbury.
Parisi, L. (2019). Media ontology and transcendental instrumentality. *Theory, Culture, Society*, 36(6), 95–124.
Pourciau, S. (2022). On the digital ocean. *Critical Inquiry*, 48(2), 233–260.
Quashie, K. (2021). *Black aliveness, or a poetics of being*. Durham, NC: Duke University Press.
Roden, D. (2014). *Posthuman life: Philosophy at the edge of the human*. New York: Routledge.
Rose, N. (2013). The human sciences in a biological age. *Theory, Culture & Society*, 30(1), 3–34.
Ruppert, E., Law, J., & Savage, M. (2013). Re-assembling social science methods: The challenge of digital devices. *Theory, Culture and Society*, 30(4), 22–46.
Sack, W. (2019). *The software arts*. Cambridge: MIT Press.
Savage, M. (2013). The 'social life of methods': A critical introduction. *Theory, Culture, Society*, 30(4), 3–21.
Savage, M., & Burrows, R. (2007). The coming crisis of empirical sociology. *Sociology*, 41(5). https://doi.org/10.1177/0038038507080443.
Serres, M. (2017). *Geometry: The third book of foundations*. (Trans. M. Randolph Burks). London: Bloomsbury Publishing.
Sholtz, J. (2021). A thousand plateaus and cosmic artisanry: On becoming destroyer of worlds. *Deleuze and Guattari Studies*, 15(2), 197–225.
Simondon, G. (1958/2017). *On the mode of existence of technical objects*. (Trans. C. Malaspina & J. Rogove). Minneapolis, MN: University of Minnesota Press.
Stengers, I. (2018). *A different science is possible: A manifesto for slow science*. Hoboken, NJ: Wiley.
Stevens, H. (2013). *Life out of sequence: A data-driven history of bioinformatics*. Chicago, IL: University of Chicago Press.
Tsing, A. L., Deger, J., Saxena, A. K., & Zhou, F. (2023). *Field guide to the patchy Anthropocene: The new nature*. Stanford, CA: Stanford University Press.
Tuhiwai Smith, L. (2021). *Decolonizing methodologies: Research and indigenous peoples*, 3rd edition. London: Bloomsbury Publishing.

Vandermere, J. (2014). *Annihilation*. New York: FSG Originals.
Wendt, A. (2015). *Quantum mind and social science: Unifying physical and social ontology*. Cambridge: Cambridge University Press.
Williamson, B. (2016). Digital education governance: Data visualization, predictive analytics, and 'real-time' policy instruments. *Journal of Education Policy*, 31(2), 123–141.
Wilson, E. (2015). *Gut feminism*. Durham, NC: Duke University Press.

2

THE METAMORPHIC MULTIPLE EARTH

Bruno Latour's Anthropocene science

Introduction

Bruno Latour was hugely instrumental in Science Studies, shaping the field in the early 1980s (Latour, 1987, 1988). Long before the "ontological turn" and the "new materialisms" (Coole & Frost, 2010), Science Study scholars in the 1980s showed how scientific practice was always engaged with the more-than-human, troubling the subject/object divide in any experiment with "consequential meaning," by engaging nonhuman "actants" with varying degrees of agency. The 2017 book *Facing Gaia: Eight Lectures on the New Climatic Regime* offers a fascinating synthesis of Latour's efforts over the last decades to find the experience at the heart of Modernity, develop an alternative to that misguided mode of being, and map a new way of worlding in the midst of our ecological crisis.[1] In focusing on the *metamorphic zone* of terrestrial life, he demands that there be a reckoning with our current climate crisis and environmental "cosmocolossus." The book dwells on the many facets of this task, distinguishing the globe from the earth, Gaia from religion, climatology and geostory from geology, and the Anthropocene from Modernity.

In this chapter, I will explore (a) the use of the term "climatic regime", (b) his emphasis on agency, animacy, and immanence as part of his "practical metaphysics," (c) his proposal for a risky terrestrial diplomacy and compositional peace under post-truth conditions, and (d) the way in which the "end times" is conceived as part of a new historicity and spatial imaginary. These topics together link his ontological commitments with his interest in the politics of knowledge-pursuing practices. Latour proposes a kind of *posthuman diplomacy* for the "end times" as a way of dealing with the increasingly tense relations between various polarized and weaponized perspectives during the Anthropocene. Such an endeavour involves a remixing of politics and earth, reckoning with the specific historical challenges we face today. In this essay, I dig into the ontological commitments underpinning Latour's image of science, and I explore the political implications of his proposal for an Anthropocene diplomacy that embraces "a Politics in which the role of Nature is explicit" (Latour, 2017, p. 47).

DOI: 10.4324/9781003538769-2

Latour's image of science is gleaned from anthropological studies of scientific activity. Through careful ant-like tracing of the actions of scientists, we can see that scientists work in a metamorphic mixture that subsumes human agency, a shapeshifting transactional zone where traits are exchanged and power is negotiated. Actor-network-theory (ANT) was meant to show how scientists have never been Modern, but have always been immersed in the relational space of multiplying agencies. Such practice is attuned to the metamorphic zone of more-than-human becoming. When scientists reject this "anthropological" image of science, because it seems to undermine claims to objectivity, they fail to realize that multiplying agency is their strength!

> The clear result of the scientific disciplines is an immense *increase* in what moves, acts, heats up, boils over, and becomes complicated – in sum, in what actually *animates* the agents that constitute the world and in the continuous refinement of the *metamorphic zone* that we encountered in the earlier lectures.
>
> *(italics in original, Latour, 2017, p. 163)*

According to Latour, science aims to uncover agencies, ensuring that claims about particular competencies on behalf of the characters involved (volcanoes, bacteria, electrons) are *entirely grounded* in the relational performances of those characters. In other words, science, as it is practiced *or as it should be practiced,* is focused on creating opportunities or experiments that reveal nonhuman agency, and thereby making visible new forms of life. Latour affirms the power of science as a practice that is committed to understanding the plurality and generativity of the earth. It is this political power of science to "represent" under-represented others, not in the way the nation state or other forms of governance do, but through a strange system of experimentation, inference, and proof, which is crucial today in the current climatic regime: "That power of representation will be of capital importance in the coming conflicts over the form of the world and the new geopolitics" (Latour, 2017, p. 33). I am always a little surprised that Latour seems so willing to engage in what seems to be a classical parliamentary approach to governance. Then again, I am also keenly aware of the necessity of this emphasis on "representation" in a struggling democracy, new or old.

In a "post-truth" era, Latour is explicit – we must defend scientific inquiry against those who contest its every claim. However, he is not defending science in the name of white cultural elitism or Euro-exceptionalism, but rather as a particular material and worlding practice that *at its best* responsibly encounters the more-than-human. This perspective affirms the capacity of science to demonstrate situated positivity, while also reckoning with the role of science in perpetrating violence of all kinds. Latour is adamant that science studies was never anti-science (as it was sometimes interpreted) but was simply affirming the *political project* of science. In other words, the material labour of science is – at its core – a political practice of forging networks. His attention to how the material labour of science shapes new kinds of societies helps us rethink "the scientific method" as being bound up with relationality and labour. For Latour, *science involves the invention of equipment (apparatus) that make perceptible (principally to humans) new forms of non-human sign-making and agency, thereby multiplying agencies across an ecology, expanding the alliances, and ultimately assembling another more-than-human political body*. Consider, for instance, the way that the human body is now known (to humans) as a quivering bacterial colony, sustained through various risky alliances and cross-species entanglements.

Thus, Science Studies for Latour is not simply a critique of false claims to objectivity, nor driven only to disclose the *relativism* and bias of scientific truth, but also aims to show *how science gains in realism* through this multiplying of agency. Unlike relativists who banish the ontological realm, he claims that the metamorphic zone "is *a property of the world itself* and not only a phenomenon of language *about* the world" (Latour, 2017, p. 69). This is not the "pseudo-realism" that purports to describe the actor (human or other) as intentionally moving about, while the context (earth or other) lies inert in the background. Latour's "realism," like Barad's agential realism, is deeply committed to the entanglement of body, place, and idea. To be entangled is more than simply recognizing the relativism of truth claims, nor is it merely a reference to the situated nature of knowledge; entanglement is about the immanent *ontology* of relationality. And it is this immanent ontology (of dependence relations) that science (at its best) pursues and elaborates.

Anthropocene, Capitalocene, Anthrobscene, Plantationocene,...

The term "Climatic regime" reminds us that "the climate" is never a pure state of nature operating outside of culture, but always populated and engendered by diverse agencies, some of them human (Verlie, 2017). Granted, the "New Regime" of the Anthropocene entails extensive power for particular humans, as is evident in the geologic record (the global distribution of plastics, etc.), and points us towards terms like Capitalocene (Moore, 2015), Anthrobscene (Parikka, 2015), and Plantationocene (Tsing, 2017). For Latour (2017), the coming ecological crisis entails "a profound mutation in our relation to the world" and a new image of humanity (p. 8). And this mutation demands new ways of "belonging to nature." In the first two book chapters, Latour revisits the notion of "nature" inherited from Anthro-centric Moderns, a notion that was used to unify the polity and "naturalize" law. Any claims to such an essentialized Nature today are more often – and rightly – contested. Latour confesses that his own contribution to this critique of "natural kinds" has doubtless played some part in assisting the adversaries and deniers of climate science, who are all too happy to cast doubts on the facts that confirm the human origin of current climate mutation (Alumkal, 2017; Drummond & Fischhoff, 2017). Notably, many of these skeptics are aligned with neoliberal and conservative religious beliefs and dismiss science as "politically motivated" if it curtails the flow of capital (Gauchat, 2012; Hamilton, 2010). Latour became concerned with these science deniers in the midst of the ecological crisis.

Indeed, the Anthropocene sets the stage for a science that must confront its Western imperialism in ways that stretch beyond reflexive critique (Avelar, 2013; Stengers, 2015, 2018). This new science might involve a kind of geohistory or geostory to serve the earthbound "terrans" who have forgotten that their planetary dwelling is also a living creature (Stengers, 2018). But the old "white geologies" encode within them an extractive logic that is precisely what has caused so much destruction. Povinelli (2016) suggests we focus on "geontologies" and resist a particular European image of humanity that fails to reckon fully with the pressing problematic of non-life (the geo). *Geontopower* characterizes that which is both at work and unravelling today, if one were to do history and politics and science as earthbound terrans rather than globalizing Moderns, and confront the "end-times" of a particular biosphere. She chooses "geo" because of its reference to "nonlife" and because it illuminates "the cramped space in which my indigenous colleagues are forced to maneuver as they attempt to keep relevant their critical analytics and practices of existence" (p. 6).

Insofar as the Anthropocene marks a period when "all human activity" has been transformed into earthly mutations, and when the climate stability of the Holocene comes to an end, it's essential to note that this moment has been manufactured by a portion of humanity invested in accelerated capitalist accumulation and white supremacy. Latour affirms the 1610 date as a possible marker for the beginning of the Anthropocene, because

> the reforestation of the American continent had, by that date, led to the stocking of so much atmospheric CO2 that climatologists could use it as a minimum quantity on the basis of which they could measure its regular increase. But why this massive reforestation? ... because of the extermination by the sword, but also by contagion and disease, of nearly fifty-four million Native Americans ....
>
> *(pp. 184–185)*

Understanding this "global" condition as the legacy of European imperialism is important, and how that legacy will also be its unraveling (Chakrabarty, 2014). There simply is no one humanity that suffers the Anthropocene, but a multiple earth of a billion black anthropocenes (Yusoff, 2018). As Chakrabarty (2012) points out:

> There is no 'humanity' that can act as a self-aware agent. The fact that the crisis of climate change will be routed through all our 'anthropological differences' can only mean that, however anthropogenic the current global warming may be in its origins, there is no corresponding 'humanity' that in its oneness can act a political agent.
>
> *(p. 15)*

These perspectives help us appreciate the differentiated life-lines in the event of the cosmocolossus. As Yusoff (2018) states:

> Seeking to monumentalize Anthropocene history is an attempt to reclaim an "innocence" around this geohistory ... To be included in the "we" of the Anthropocene is to be silenced by a claim to universalism that fails to notice its subjugations ....
>
> *(Yusoff, 2018, pp. 11–12)*

Despite these concerns, I am all too conscious that computational culture writ large has colonized the earth. There is now a disturbing digital life that seems to occupy the globe – the digital is both what seems to connect these different localisms and, at the same time, be a platform for spreading pan-human stupidity (de Freitas, 2018). Latour turned to digital ethnography in the 2000s to track this connectivity across online platforms, by partnering with the statistician Lépinay, and drawing on ideas from the 19th century posthumanist Gabrielle Tarde (influential for Deleuze and Guattari as well) who argued for alternative ways of mobilizing quantitative methods (Latour, 2002). Tarde was a 1890 French writer who argued for a statistics that was less about forcing data to comply to pre-given models, and more about letting the data-vectors of miniscule everyday acts accumulate in big data sets from which models of behaviour and sociality emerge (I discuss this in more detail in Chapter 3). AI today is a big data enterprise, and new empiricisms are emerging alongside these habits of inquiry, as the earth becomes a massive digital sensor, cloaked in data clouds (Gabrys, 2016).

There is an urgency to Latour's tone throughout *Gaia*, a noticeable anxiety to his claims, as he dwells on the eco-politics and onto-epistemologies of "The ends of the world" (Danowski & Viveiros de Castro, 2017). All too aware that the Eurocentric vision of humanity is not how the diverse people of the earth will undergo the Anthropocene, he is focused on "terrans" as earth dwellers of all kinds. Concerned that many are moving along like sleepwalkers, oblivious to the destruction writ by the flow of capital, Latour asks that his readers consider our ecological crisis as "a generalized state of war" (Latour, 2017, p. 73). He is not advocating for the kind of geo-engineering solution that aims to defeat an enemy combatant but is concerned with a passive melancholia that causes many to hide from the horrors of the day. Rather than speak of "hope" (which is always haunted with a desire for the past) or passive despair, he asks that his reader discover new "forms" of the earth and different ways of experiencing the passage of time. He asks us to experiment and think speculatively, to forge alliances across art and science, and to care for the future. Latour's arguments against human exceptionalism are always grounded in particular proposals for action, as he aims to give humans "a shape" or "morph humans into a more realistic image" (Latour, 2017, p. 110).

Agency, animacy, and causality

What becomes of causality in this approach to research? How might social science research investigate relationships where the agency is distributed across a complex hybrid ecology? Latour argues that conventional renderings of science rest on a faulty concept of causality, which ascribes all the agency to a deterministic cause. The conventional "scientific worldview" has often been used reductively to *deanimate* matter, and *overanimate* human intellect as both the cause of activity and the diviner of causes. This image of science fails to accurately characterize *the practice* of science and reflects instead a Modernist desire to dominate and colonize. This "causalist" approach keeps all the action in the "stimulus," as though action was always sourced from a unity, with the result of deanimating that which undergoes the intervention. This is the fantasy of control – the desire that there will be no surprises, no feedback loops, no reciprocal touch. By deanimating the non-human and overanimating the human, the Moderns fail to encounter an earth populated with animated agents (Haraway, 2016). Worse still, this approach de-dramatizes the unfolding of events and the course of time, so that the future is made entirely predictable and subjected to the force of the "present." To buck the causalist plot is to live in a world where future agents (typically demoted as *consequences* and *effects*) signify and bring their existence retroactively into the present. Shedding our attachment to causes that literally control all the action would entail queering time as we know it. This demands a new kind of causality which tracks the reciprocal mutations of the metamorphic zone, linking animacies and agencies *across times* in new ways.

Such thinking runs counter to conventional images of the scientific method, which remain tied to a Newtonian notion of cause-effect. If the Western scientific revolution begins with Galileo who exclaims, against the ontology and dogma of his day, that the earth moves: "Eppur si muove" ("and yet it moves"), today the West confronts an earth that is not only a moving sphere amongst many, but also an "active, local, limited, sensitive, fragile, trembling, and easily irritated envelope" (Latour, 2017, p. 60).[2] *Galileo's earth had movement, but no behaviour nor life itself*. The earth known as Gaia is instead fully "situated" (Haraway, 1988), and "inflicts" on humans their return from the infinite realm of ideals back into the folds of the material finite world. Gaia has two fundamental features: *it is made of agents that are neither deanimated nor other-animated, and it is made of agents that are not prematurely unified in a single acting whole*. Indeed, the *composition* of Gaia is profoundly unknown, shaped by an infinity of heterogeneous agencies and waves of action that have the potential to defy globalism. This is Latour's multiple earth.

Science must *not deanimate* the earth, but instead treat it as a being in disequilibrium, and attend to its colour, smell, surface, texture, aging (death), etc. Gaia is not a figure of harmony. Her mythic origins portray her as a violent trickster. Latour inherits the figure of Gaia from James Lovelock and microbiologist Lynn Margolis, and their image of earth as a complex system and organism. This reliance on the ideas of Lovelock, however, raises the concern that the entire project operates under the confining concept of *organism*. Lovelock (1979/2000) described the earth as a "super-organism" such that "the earth behaves as a self-regulating system, and … the proper science for its study is physiology." Although Latour argues that Gaia outstrips any biological image of thought (and body), I am still wary of how the concept of organism drives this image of the earth. Gaian bio-geopolitics has had huge impact on systems theory, or what Clarke (2017) calls "systems counterculture" through the work of Francisco Varela and William Irwin Thompson, and continues to echo in the work of Esposito (2008, 2011) on biopolitics and immunity. But perhaps we need to consider quantum paradigms and other scientific shifts, during the Anthropocene, in our attempts to understand how a body – any body, be it an atom, human or planet – is formed, organized, and coordinated under some other form or image of life, different from the organism. Karen Barad's work on the quantum paradigm is relevant here, as we reconsider causality and relationality in 21st century scientific theories (Barad 2012a, 2012b). Moreover, I think there is a need to think beyond the conventions of organicism, and examine the way that digital methods raise questions of quantum life and quantum decision theory (for more see Chapter 4 and also de Freitas & Sinclair, 2018).

The Anthropocene calls for a form of inquiry and engagement that is truly *immanent*. This would be a kind of immanence that would bring causality down into the mud of effects and consequences, a feedback situation that might better reckon with human activity in the flesh. Immanence fuses the finite with the infinite, but not as theological *eternity* in finite time, found in the religious doctrine of incarnation, and articulated in terms of an *emanation of god in matter*. Instead, Latour argues that *immanence must be cause of itself*.[3] Reclaiming causality to be of the world and in the world (as cause of itself) demands that researchers rethink the event-nature of their research methods. The proposal that substance is "cause of itself" has a long history and is linked to the ideas of Spinoza in Western philosophy (de Freitas, 2018).

Latour describes immanence in contrast to *emanation*. Emanation is linked to the dangerous idea of "immanentization of the transcendental," often found in the rhetoric of *incarnation*. The danger here is that incarnation is an act of immanentization that inserts the transcendent into the material as though saving the profane body – *the material world is then recuperated, rescued for having a right to life, and then subject to pursuant crimes in the name of liberation and enlightenment*. This kind of incarnation actually betrays the very idea of immanence. Moreover, the ruling affect is regret rather than joy. Dejection and melancholia come from not being able to think the end times within a history that carries on, not being able to accept a geostory that has multiple futures and multiple earths. Joy is the affective correlate of immanence. Beware those who deny this potential multiplicity as inhering in the world; beware even those who claim to grasp a spirit which *emanates from matter* if they also refuse to imagine a posthuman eternity. It's important that immanence *not* bear eternal definitive human value, but rather describe the eternal *incorporation* of a pluralist, dynamic, and open set of agencies that *inhere* in the world.

Convoked authorities and posthuman diplomacies

Latour's image of political ecology and posthuman science can help us engage with contemporary polarized positions on human rights, while we witness a return to anti-immigration nationalisms and religious fundamentalisms. Latour's *Gaia* is based on a prestigious lecture series hosted in Edinburgh on "Natural Religion," delivered in the past by the likes of Alfred Whitehead (in 1927) and Niels Bohr (in 1949). The lectures are an attempt to bring secular scientists together into diplomatic relations with other "convoked authorities" that go by some other name (i.e. religions). This approach frames "collectives" more generally, according to that which is cared for (either fact or fiction). Achieved by way of its *rituals*, a collected people is *convoked* around that for which they care. Thus, a collective or people or body will form around, for instance, a practice such as scientific inquiry, but also around a religious set of beliefs. This is not to say that science is simply another form of religion, but rather to recognize the pluralism of collective human practices as they rigidify.

Latour draws on the historian of religion Jan Assmann (2010, 2018) who recounts how prior to monotheism, there were mosaic cosmopolitan communities across the Mediterranean territories. Rather than insist on a deity that was untranslatable and incommensurable with others encountered, translation tables listed attributes, so that one could describe one's own passionate attachments in such a way that the other might begin to recognize their own. Your god is called X and mine Y, but they might share some attributes (trickster, storm-maker, omniscient, etc.). Latour advocates for a similar space of encounter and translation as a way to co-exist and better understand how people differ. This early kind of mosaic culture was pluralistic because it thrived on a certain kind of engagement that ensured linkage or looping across borders, beliefs, and territories, but Latour doesn't adequately address the power differentials across these loops, nor the fact that slave labour and extreme gender restrictions made for radical inequality. Colebrook (2017) critiques this kind of all-consuming translation, whereby everything comes to relate to everything else. What gets lost in translation?

And yet I admire Latour's concerns that "faith" cultures are dangerous. Granting oracle status to a being of any kind, including an AI algorithm, is abnegating agency to a mysterious and cloaked authority. Latour critiques the relation of "faith" as an inadequate engagement with the world. Referencing monotheistic religions, he sees faith as a dangerous construct, since it must pursue spirit not through the assurances of the body and the encounter, but through a fragile, uncertain gesture without ground (and typically through indirect priestly interpretation as well). He contrasts these kinds of religions with philosophies of immanence, where engagement and encounter are privileged over indirect representation. Faith people live in an epoch after the end of time, but engage their worldly gods and forces through mediated distance and/or salvation. This demands a tenuous communing with the real, since the real is lost in its actuality. The Moderns inherit the habit of faith, a detached relationship to the earth: "*they have no sort of possible contact with the terrestrial*, since they can see in it only the transcendent, which would be trying awkwardly to fold itself into the immanent" (Latour, 2017, p. 205).

This term terrestrial is also crucial for setting aside the flawed relationships with the earth that we've inherited from the "rational" Moderns. For Latour (2017), "the terrestrial is *immanence freed of immanentization*" (p. 212), which thereby avoids the dualism of mind and matter. For Kaiser and Thiele (2017), however, the term proves powerful in directing our attention to at least three aspects of this new order:

> terrestrial as tool to think human existence emancipated from a divine order; terrestrial as tied to earth in view of the immanent realm of planetary existence; and terrestrial in the sf-mode as envisioning terran existences – as earthly critters, and thereby disrupting the structural verticality of heaven and earth and the anthropocentric fantasies of extraterrestrialism.
>
> *(p. 427)*

Latour proposes five key questions that must be posed to any creature, no matter their status or location on earth, questions that must be posed and answered. As much as these are questions meant for humans with contesting beliefs, for me, they evoke a kind of interstellar diplomacy, and I can imagine them articulated in a StarTrek episode. Any form of risky diplomacy amongst creatures naming themselves as human (however they define that), or fighting to occupy and share the earth, must begin with these questions and requests: (1) identify the "supreme authority" by which you believe you have been convoked (i.e. military power, god, scientific knowledge, sexuality, nation state, etc.), (2) describe what limits you are given as members of this group, (3) state what territory you believe you inhabit, (4) name the epoch in which you believe you live, and (5) explain the principle of organization that distributes agency across your networks. Latour suggests that posthuman diplomacy be pursued through some such exercise of posthuman encountering, where competing parties might create a conversation that crosses radical divides, and where science and religion might find a way to communicate about their differences diplomatically. Perhaps the cosmic artisan should carry this list of questions with them as they go about their investigations.

Question one emphasizes the need to name your *authority* and offers an important transparency for any diplomacy to begin. Question two identifies the *limits* that curtail activity and thus makes transparent the existence of any rule of law, and any responsibilities that a community or collective must respect. This emphasis on our embodied and collective limitations is precisely what helps us understand our relations of dependence. Questions three and four reference territory and epoch, drawing attention to how two people, initially unknown to each other, come to understand how they conceptualize their borders and their rights to "occupy" a place at a given time, and also whether they think they are starting out, at the beginning of their formation, or perhaps are a self-described "chosen people" that come "after" some previous societal fall. If answered honestly (a big "if"), these questions might open up an opportunity for translation and understanding across radical differences. The proposal for diplomatic transparency seems a tad rationalist in its trust in perfect translation, but the emphasis on power and limitation is notable. As Latour (2017) states repeatedly, "The earthbound have to explore the question of their limits" (pp. 290–291). He emphasizes the finitude of the planet and of life: "*To live in the time of the end* is first of all to accept the finitude of the time that passes and to put an end to negligence" (Latour, 2017, p. 286). Sensitive to how idealisms have caused so much harm in the name of colonial salvation, Latour announces: "To rediscover meaning in the question of emancipation, *we have to free ourselves from the infinite*" (Latour, 2017, p. 285). It's important to note that this "infinite" is the ideal extracted infinite, that is to say, the ideal infinite that transcends time and place, and not the *immanent* infinite that occupies the here and now, as we find in Spinoza and Leibniz (see de Freitas, 2018, for more).

Time and space: flows, spheres, and loops

So much hinges on the way people conceptualize time, including their relationship to the future. *What epoch do you believe you are in?* How does your understanding of time impact the way you relate to the earth and others? Do you believe that you are in an indefinite present to which all other times lead? Or a time after "the fall," where you must seek forgiveness? Or a time that feels the force of an unruly and unknown future feedback loop? Latour notes that talk of the Anthropocene brings up talk of *apocalypse* and end times, and perhaps even a desire for *the end of time*. Moreover, "there is no doubt about this point: the West has landed on all other civilizations like an Apocalypse that has put an end to their existence" (Latour, 2017, p. 205). With total calm the Moderns inflict apocalypse on others, as though they had a right to "the" earth. This kind of entitlement betrays the fact that the earth they claim is emptied of animacy, and therefore without adequate temporality. Ironically then, the earth is no longer terrestrial (spatial and temporal) in the Modernist fantasy of control: "there is nothing terrestrial about this Earth, since what is denied, precisely, is that it has a history, a historicity, a retroaction, capacities – in short, agency" (Latour, 2017, p. 207).

For many dominant world religions today, we are already *after the end*, reprising and operating under orders of salvation. Under such conditions, ecological warnings go unheeded. "To these peoples, obviously, nothing serious can happen any longer, since they believe they have always been within the 'end of history'" (Latour, 2017, p. 195). These are the moralists who respond to climate science condescendingly that they have already passed to the other side, no longer of our world, having entered into the kingdom of God or some other space of detachment (which could be righteous or apathetic). They shrug and say that ecological disaster and the end of life on earth is part of the plan. For these people time has already ended, and there is no longer any meddling in temporal flow (no sharing in the metamorphic zone). This makes the earth without history, shielded from the passing of time. These are people who are perhaps the best at negligence – always looking back or indirectly at the world, rather than engaged with it.

To fight this rampant negligence, the earthbound must reckon with "an end of time within time" (Latour, 2017, p. 194). This is an end that is encompassed in the world, because there is a "time of the end" that must be attended, but not a universal and singular "end of times" (Latour, 2017, p. 197). The earthbound are always recommencing, always uncertain about time's achievement, but equally certain about its historical flow. It is this awkward sense of history – paradoxical in some sense – that must be kept alive. To counter negligence and apathy, we must reckon with the limits of our science and religion, to ensure that these attachments do not surrender us to a higher detached temporal order (a transcendent realm). Life after death must remain entirely terrestrial.[4] In other words, the Moderns (be they bad transcendentalists of the scientific or religious kind) have broken the crucial bond with temporal historical flow. The end, for Latour, is an achievement, a finitude, and only a revelation insofar as it is *in and with* time, "and especially with the passage of time as its necessary medium" (Latour, 2017, p. 197). He will offer a strange maxim: "What lasts forever lasts only through what does not last" (Latour, 2017, p. 197).

The term *intensity* is key for Latour as he reimagines the earth as a network rather than a sphere or globe: "We are still dealing with space, with the earth, with discovery, but it is the discovery of a new Earth considered in its *intensity* and no longer in its *extension*" (Latour, 2017, p. 290). Following Sloterdijk's (2011, 2016) work on Modernity, globalism, and spheres, Latour critiques the form of the sphere in Enlightenment images of the planet. The sphere itself has constant curvature, and *performs* rounded, continuous, complete, perfectly symmetric, and *traversable knowledge*. This rendering masks the actual material practices of assembling such a spherical form, and it overcodes the multiple earths that exist.

Moderns adore the perfect sphere: "a sphere has no history, no beginning, no end, no holes, no discontinuities of any sort. It is not merely an idea but the very ideal of ideas" (Latour, 2017, p. 136). It is in the age of European expansion and colonial acquisition, when terrestrial navigation spreads the monosphere across the earth, and a "vulgar" and diminished globalization takes hold. It is this figure of the globe with a capital G that must be interrogated, for how it is built, tended, traversed, inhabited, consumed, etc. Latour suggests that building a globe, "always amounts to reactivating a theological theme" whereby aims for encyclopedic knowledge, transcendence, being the chosen people, etc., are given a popular form. Latour continues, claiming that the danger of spheres is that they both demand and enable a "premature leap to a higher level *by confusing the figures of connection with those of totality*" (Latour, 2017, p. 130). I will revisit this complex concept of totality later in Chapter 13, and reconsider its potential in the work of Édouard Glissant, who embraces an archipelagic totality. I also discuss some critiques of Latour's bottom-up approach to totality – drawing on Wark (2017) and others – in chapter 11.

The concept of the loop is central in the work of Latour, as part of his network-image of more-than-human society. In his emphasis on open wholes and partial affiliations, loops play a significant role:

> ... we have to slip into, envelop ourselves within, a large number of loops, so that, gradually, step by step, knowledge of the place in which we live and of the requirements of our atmospheric condition can gain greater pertinence and be experienced as urgent. The slow operation that consists in being enveloped in sensor circuits in the form of loops: this is what is meant by "being of this Earth."
>
> *(Latour, 2017, p. 139)*

With each passage through a loop, one increases sensitivity to the fragility of the multiple earth. Moving through loops is a way to refuse the global perspective, and remain down in the tangles – the job is to deploy instruments that further allow us to make visible and sensible these paths, and to detect these connections between diverse agencies. This increased sensitivity is about "detecting and responding rapidly to small changes" (Latour, 2017, p. 141). We witness here a new kind of "reflexivity" that brings with it an intensity, "woven loop after loop, reflexivity after reflexivity, instrument after instrument" (Latour, 2017, p. 143). Ecologists have used the concept of the loop to explore the twisted and disorienting nature of connectivity (Morton, 2016).

A focus on loops is also found in Latour's earlier work when he redefined sociology, not as "the science of the social" but as "the tracing of associations" in ANT (Latour, 2005, p. 5). In many ways, Latour remains, as Graham Harman (2010) said, "The prince of networks." He coined the term Actor Network Theory in the 1980s, before the internet had taken off, and before the term network came to be used for everything, watering down its meaning to the point where he became somewhat ambivalent about using the tag, proposing instead *work-net theory*, to emphasize the labour involved in sustaining a network. Other titular candidates at the time of inception had been "meshwork" and "network of innovation" and "actant-rhyzome ontology," indicating his indebtedness to the ideas of Deleuze and Guattari (1987). These terms also speak more accurately to his later work on the complex computational meshwork that binds contemporary scientists to their subject matter (see Chapter 4).

Last words: territory, conflict, and the digital earth

Research today is often under the gun, and seen as an elite indulgence that fails to serve the larger public. In his powerful 2004 essay *Why has critique run out of steam? From matters of fact to matters of concern*, Latour anticipated our current crisis in academia, where identity politics and battles over language have distracted us from more significant cultural and material transformations:

> Wars. So many wars. Wars outside and wars inside. Cultural wars, science wars, and wars against terrorism. Wars against poverty and wars against the poor. Wars against ignorance and wars out of ignorance. My question is simple: Should we be at war, too, we, the scholars, the intellectuals? Is it really our duty to add fresh ruins to fields of ruins? Is it really the task of the humanities to add deconstruction to destruction? More iconoclasm to iconoclasm? What has become of the critical spirit? Has it run out of steam? Quite simply, my worry is that it might not be aiming at the right target.
>
> *(Latour, 2004, p. 225)*

His essay targets those scholars battling each other across the humanities, in the name of critique, when there might be a more important battle in which they should engage. In his later work, Latour (2018) continues to voice this concern, while the global corporate "elite" are managing this war – allowing destruction and violence to carry on – achieved through a kind of globalization that reigns over all human conflicts. This particular brand of economic globalism effectively *depoliticizes ecology* and *deanimates the earth*. Those of us who are comfortably rewarded by this arrangement carry on like sleepwalkers, oblivious to the destruction. Under this regime, the earth is sacrificed. How does one climb out of a control state, rising up against the quiet policing that keeps this regime intact? In the current sleep-walking condition, indifferent to the ecological disasters in which we participate, desires and delusions continue to subject the deanimated earth, at global scales.

A repoliticization of ecology involves recognizing the antagonistic relationality in which we are invested, and the need to reshape dependency relations.[5] Gaia is the multiplication of sites where agencies proliferate, where conflicts and contested territory are recognized and diplomatically engaged. Latour asks that we acknowledge there has been an ongoing war, and that there is no pure harmony with nature nor with each other. He pursues a posthuman diplomacy as the alternative to war, a process of declaring limitations and dependencies, using the interview/encounter protocol discussed above:

> Bringing out these limits is the only way to repoliticize ecology and to put an end, consequently, to the simple operations of conquest, land grabs, or pacification.
>
> *(Latour, 2017, p. 241)*

New digital wrappings of the planet and the saturation of everyday life with computational AI technologies will have a hand in this new diplomacy (Gabrys, 2016). The earthbound must realize that the control state is a generalized state of war, in which negligence reigns. Such negligence often goes hand-in-hand with assumptions that the earth will take care of itself. Gilbert et al. (2012) contest the very idea of an earth that is a "host" in a feedback loop, an image that fuels current neocybernetic images of the planet as home or mother; Clarke (2017) explains how somatic individuation (and the practice of immunity that defends an organism or a crown) is itself a cosmological evolved state of a more general entropic dispersion of matter. The earth is perhaps entirely indifferent to its own existence, let alone ours, and we might need to come to terms with that indifference. Thinking at these planetary scales is necessary, as we immerse ourselves in a global computational culture, where AI network technologies enact cybernetic fantasies of negentropic control.

Notes

1. The 2018 *Down to Earth: Politics in the New Climatic Regime* is not quite as powerful, although it offers insight into the same content and argument.
2. Latour points out that the account of "Galileo the renegade" is now used by climate skeptics when they attack the apparent consensus of climatologists.
3. The Spinozan emphasis on substance as "cause of itself" captures the kind of philosophy of immanence characterized here.
4. I am conscious that this paper puts challenges on "me," as to choosing particular pronouns and forms of address. Latour uses "we" often, purposely conjuring a collective. That collective is perhaps better characterized by "the earthbound."
5. Latour suggests in his 2018 book *Down to Earth* that we might need to stop using the term "ecology," since ecology as a form of inquiry has in some fundamental sense failed to shift the paradigm, the way Karl Polanyi predicted in 1945 in *The Great Transformation*.

References

Alumkal, A. (2017). *Paranoid science: The Christian right's war on reality*. New York: New York University Press.

Assmann, J. (2010). *The price of monotheism*. (Trans. R. Savage). Stanford, CA: Stanford University Press.

Assmann, J. (2018). *The invention of religion: Faith and covenant in the book of exodus*. Princeton, NJ: Princeton University Press.

Avelar, I. (2013). Amerindian perspectivism and non-human rights. *After/Nativas*, 1, 1–21. Available at https://alternativas.osu.edu/en/issues/autumn-2013/essays/avelar.html.

Barad, K. (2012a). On touching: The inhuman that therefore I am. *Differences: A Journal of Feminist Cultural Studies*, 23(3), 206–223. Revised and republished in *Power of Material/Politics of Materiality* edited by Susanne Witzgall and Kirsten Stakemeier, 153–165.

Barad, K. (2012b). Nature's queer performativity. *Women, Gender and Research*, 1–2, 25–53.

Chakrabarty, D, (2012). Postcolonial studies and the challenge of climate change. *New Literary History*, 32(1), 1–18.

Chakrabarty, D. (2014). Climate and capitalism: On conjoined histories. *Critical Inquiry*, 42, 1–23.

Clarke, B. (2017). Planetary immunity: Biopolitics, Gaia theory, the holobiont, and the systems counterculture. In E. Horl & J. Burton (Eds.), *General ecology: The new ecological paradigm* (pp. 193–216). London: Bloomsbury Press.

Colebrook, C. (2017). *Essays on extinction*. Open Humanities Press. Creative Commons. Available at https://www.openhumanitiespress.org/books/titles/sex-after-life.

Coole, D., & Frost, S. (2010). *New materialisms: Ontology, agency, and politics*. Durham, NC: Duke University Press.

Danowski, D. & Viveiros de Castro, E. (2017). *The ends of the world*. (Trans. R. Nunes). Cambridge: Polity Press.

de Freitas, E. (2018). Love of learning: Amorous and fatal. In R. Braidotti & S. Bignalle (Eds.), *Posthuman ecologies: Complexity and process after Deleuze*. London: Rowman & Littlefield.

de Freitas, E., & Sinclair, N. (2018). The quantum mind: Alternative ways of reasoning with uncertainty. *Canadian Journal of Science, Mathematics and Technology Education*, 18(3), 271–283.

Deleuze, G., & Guattari, F. (1987). *A thousand plateaus: Capitalism and schizophrenia*. (Trans. B. Massumi). Minneapolis: University of Minnesota Press.

Drummond, C., & Fischhoff, B. (2017). Individuals with greater science literacy and education have more polarized beliefs on controversial science topics. *Proceedings of the National Academy of Sciences*, 114(36), 9587–9592. https://doi.org/10.1073/pnas.1704882114.

Esposito, R. (2008). *Bios: Biopolitics and philosophy*. (Trans. T. Campbell). Minneapolis, MN: University of Minnesota Press.

Esposito, R. (2011). *Immunitas: The protection and negation of life*. (Trans. Z. Hanafi). Malden, MA: Polity Press.

Gabrys, J. (2016). *Program earth: Environmental sensing, technology and the making of a computational planet*. Minneapolis: University of Minnesota Press.

Gauchat, G. (2012). Politicization of science in the ublic sphere: A study of public trust in the United States, 1974 to 2010. *American Sociological Review*, 77(2), 167–187.

Gilbert, S. F., Sapp, J., & Tauber, A. I. (2012). A symbiotic view of life: We have never been individuals. *The Quarterly Review of Biology*, 87(4), 325–41.

Hamilton, C. (2010). *Requeim for a species: Why we resist the truth about climate change*. New York: Routledge.

Haraway, D. J. (1988). Situated knowledges: The science question in Feminism and the privilege of partial perspective. *Feminist Studies*, 14(3), 575–599.

Haraway, D. J. (2016). *Staying with the trouble: Making kin in the Chthulucene*. Durham, NC: Duke University Press.

Harman, G. (2010). *Towards speculative realism: Essays and lectures*. Winchester: Zero Books.

Kaiser, B. M., & Thiele, K. (2017). Terrestrial. In R. Braidotti & M. Hlavajova (Eds.), *Posthuman glossary* (pp. 427–428). Bloomsbury Academic.

Latour, B. (1987). *Science in action: How to follow scientists and engineers through society*. Cambridge, MA: Harvard University Press.

Latour, B. (1988). *The pasteurization of France*. (Trans. A. Sheridan & J. Law). Cambridge, MA: Harvard University Press.

Latour, B. (2002). Gabriele Tarde and the end of the social. In P. Joyce (Ed.), *The social in question: New bearings in history and the social sciences* (pp. 117–132). London: Routledge.
Latour, B. (2004). Why has critique run out of steam? From matters of fact to matters of concern. *Critical Inquiry*, 30, 225–248.
Latour, B. (2005). *Reassembling the social: An introduction to actor-network-theory*. Oxford: Oxford University Press.
Latour, B. (2010). Tarde's idea of quantification. In M. Candea (Ed.), *The social after Gabrielle Tarde: Debates and assessments* (pp. 147–164). New York: Routledge.
Latour, B. (2012). The whole is always bigger than its parts – A digital test of Gabrielle Tarde's monads. *The British Journal of Sociology*, 63(4), 590–615.
Latour, B (2017). *Facing Gaia: Eight lectures on the new climatic regime*. Cambridge: Polity Press.
Latour, B. (2018). *Down to earth: Politics in the new climatic regime*. Cambridge: Polity Press.
Lovelock, J. (1979/2000). *Gaia: A new look at life on earth*, 3rd edition. Oxford: Oxford University Press.
Moore, J. W. (2015). *Capitalism in the web of life*. Brooklyn, NY: Verso Books.
Morton, T. (2016). *Dark ecology*. NY: Columbia University Press.
Parikka, J. (2015). *The anthrobscene*. Minneapolis, MN: Minnesota University Press.
Povinelli, E. A. (2016). *Geontologies: A requiem for late liberalism*. Durham, NJ: Duke University Press.
Schmitt, C. (2006). *The nomos of the earth in the international law of jus publicum Europaeum*. (Trans. G. L. Ulmen). New York: Telus Press Publishing.
Schmitt, C. (2011). *Writings on war*. (Trans. T. Nunan). Cambridge: Polity Press.
Sloterdijk, P. (2011). *Bubbles: Spheres volume I*. Los Angeles: Semiotexte.
Sloterdijk, P. (2016). *Foam: Spheres volume III*. Los Angeles: Semiotexte.
Stengers, I. (2015). *In catastrophic times: Resisting the coming barbarism*. (Trans. A. Goffey). Paris: Open Humanities Press/Meson Press.
Stengers, I. (2018). *Another science is possible: A manifesto for slow science*. London: Polity Press.
Tsing, A. (2017). *The mushroom at the end of the world: On the possibility of life in capitalist ruins*. Princeton, NJ: Princeton University Press.
Verlie, B. (2017). Rethinking climate education: Climate as entanglement. *Educational Studies: A Journal of the American Educational Studies Association*, 53(6), 560–572.
Wark, M. (2017). Bruno Latour: Occupy earth. Available at https://www.versobooks.com/blogs/3425-bruno-latour-occupy-earth (accessed 25 February 2019).
Yusoff, K. (2018). *A billion black Anthropocenes or none*. Minneapolis, MN: University of Minnesota Press.

3

NEW EMPIRICISMS OF THE FRACTAL FOLD

Rethinking monadology in digital times

Introduction

New developments in digital technologies, software analytics, and computational power are changing the relationship between the quantitative and the qualitative in scientific methods. These developments have occurred alongside a growing interest in philosophies of immanence and a widespread turn to the study of non-human agency (Bennett, 2010; Braidotti, 2013; Coole & Frost, 2010). This confluence of ideas and practices has changed the way we think about the relationship between mind and matter, with implications for empiricism in the social sciences (Sorensen, 2011). A new configuration of social science is emerging in these digital times, as we tap into new kinds of data that trouble conventions regarding what constitutes the *unit* of analysis, and question the extent to which this data is owned or even correlated to a definitive organic and individuated subject customarily referred to as "human." These methodological shifts demand a more careful consideration of the historical lineage of empiricism and its relation to the history of science more generally. Such historical work can help identify and articulate the specific contributions of new empiricisms.

In this chapter, I track the historical mutations of *monadology,* a philosophical approach to empiricism well suited to these digital times. The term monadology refers to a metaphysical theory formally developed by the 18th century philosopher Gottfried Leibniz, who rejected Cartesian dualist ontology. Leibniz argued that mind and matter were composed of the same simple substance, that being the monad. Although "simple," each monad expressed the universe and was thereby composed of infinite folds. Leibniz is likely to have taken the term "monad" from the tradition of Pythagorean philosophy, and perhaps also from other scholars who had used it in related but different ways (Strickland, 2014). Although not *new*, Leibniz's monadology text, published in 1714 and comprising less than 6,000 words, has inspired a vast array of diverse proposals for how mind and matter commingle.

Both Gilles Deleuze and Bruno Latour ascribe to variants of monadology in their proposals for a new empiricism. In the case of Deleuze (1988a, 1988b, 1993), monadology

DOI: 10.4324/9781003538769-3

is essential for his theory of difference and multiplicity, and for how he studies the virtual dimensions of matter. In the case of Latour (2010, 2012), monadology becomes a way of rethinking how degrees of agency are distributed across human and non-human agents. Both Deleuze and Latour will draw extensively on the work of Gabriel Tarde (1843–1904), a French sociologist, judge, and author of the audacious posthumanist 1895 text *Sociology and monadology*. Unlike Leibniz, Tarde proposed an atheistic monadology, in which monads aggregate because of desire and avidity, rather than teleology or destiny. Tarde uses the monadology as a way to rethink social science methods, avoiding the binary between agent and structure, and developing an alternative way of thinking about quantitative methods in the social sciences.

In this chapter, I discuss how monadology helps us rethink research methods in these digital times. Drawing primarily from Deleuze and Latour, I argue that a *fractal monadology* reassembles the fold with the digital, the continuous with the discrete, and ultimately offers a philosophical foundation for contemporary social science research. In so doing, I follow other work in related fields that seeks to develop an adequate notion of the *recombinant subject* that is not identified with a numerical statistical data point, but with fractal complexity (Galloway, 2014; Massumi, 1992; Portanova, 2009). More recently, post-colonial thinkers have turned to fractal thinking as a way of making sense of racialized and intersectional experience (Da Silva, 2016; Eglash, 2023). The fractal – as a fractional dimension - troubles our assumptions about structure, embodiment, and spatial sensibility. Fractal geometries draw attention to how macro-structures are engendered by iterative and recursive self-similar micro-tendencies unfolding at arbitrarily small scales, which seems a trope well suited to our current algorithmic conditions.

The fold

Initially, one might think that the continuous fold is simply at odds with the digital, since the two seem to reference two fundamentally distinct modes of being. But in a world where the flows of desire and affiliation are increasingly digitized and quantified at scales beyond human perception, where we gleefully participate in calculated publics and software cultures, and 44where neuroscience shows that we often perform calculations unconsciously, it seems rather urgent that we reconsider the nature of the fold:

> We are discovering new ways of folding, akin to new envelopments, but we all remain Leibnizan because what always matters is folding, unfolding, refolding.
>
> *(Deleuze, 1993, p. 137)*

Deleuze (1993) characterizes monadology as a way of theorizing multiplicity and difference as fundamental ontological forces. Monadology offers an alternative way of thinking about individuation and distinctness, recasting the relationship between the one and the many. Distinctness is no longer that which separates and cuts off one individual or object from another, but refers rather to a particular fold or twist in the undulating fabric of the universe. Processes of individuation, by which identities and subjects and institutions come into being, are not acts of disconnection or separation, whereby the one is cut off from the rest, but are continuous topological folds of the whole. Thus, a monadology works through topological concepts of connectivity and elasticity, in which individuals emerge through continuous stretching and distortion. This process describes all being, whether it be the birth of an idea or concept, or the making of a bowl. If we were able to see monads, we would see how bowls and concepts were simply folds of the one simple substance. This is a haptic theory of contiguous relationality, a way of studying life as it contracts and expands across a continuum of mind-matter. The smallest unit of matter or life, thus, is not the atom or any other particle, but is rather the fold. Leibniz offers a truly relational ontology:

> The division of the continuous must not be taken as of sand dividing into grains, but as that of a sheet of paper or of a tunic in folds, in such a way that an infinite number of folds can be produced, some smaller than others, but without the body ever dissolving into points or minima.
>
> *(Leibniz, Pacidus Philalethi (C, 614–615), in Deleuze, 1993, p. 6)*

Thus Leibniz's monads are riddled with folds that produce caverns within caverns, "each" compressed differently by the forces of appetite and perception. For Leibniz, each monad expressed the entirety of the universe, a distinct perspective or contraction of the relational forces that saturate the monadology. A fold is never final, never a definitive cut, and thus the mechanism of creation is invagination, pleating, further folding and twisting, rather than unfolding or cutting. For Leibniz, this "muscular conception of matter" corresponds to a spirit in matter, a vitalism or virtuality inherent to matter itself (Deleuze, 1993, p. 7).

This focus on the fold is related to Leibniz's development of the concept of the infinitesimal as the "smallest interval," an interval that is smaller than any conceivable measurement, but still, and yet paradoxically, not zero. Deleuze (1994) will use this concept to argue for the idea of *difference in itself* and claim that we must study *degrees of difference* rather than *differences in degree*. In other words, the fold, as the basic ontological "unit," captures this idea of difference in itself because it can always be further folded. Two distinct individuals partake in this difference in differing degrees (some more folded than others), but there is no absolute measure that will allow us to separate them, since folding is mutually transitive. Thus, the emphasis on the monadological fold rebukes a difference (or distinctness/individuation) that is determined through reference to something outside of it. With this emphasis, we begin to glimpse the consequences for empiricism, since the monadology demands that we shift from "interaction" between distinct individuals, to intra-action, much like the proposal of Karen Barad to rethink relational ontology using ideas from quantum physics (2007).

Between the years 1886 and 1904, Gabriele Tarde develops these ideas within the field of sociology, arguing that there is never a complete cut-out of the individual, only folds and knotted twists in the one flowing substance of society. For Tarde, everything is a society, including cells, viruses, ants, rocks, and schools. Each society is a particular configuration of the one simple substance – the monad – whether it be human or non-human. The key for Tarde is that difference and variation underlie all illusions of oneness and unity, so that *difference in itself* – rather than difference between distinct individuals – is inherent to all matter and life. Rather than Leibniz's appetite and perception, Tarde's monadology folds according to the two fundamental actions of *imitation* and *invention* (much like Deleuze's repetition and difference). Invention for Tarde is not the neo-liberal celebration of the entrepreneur. Tarde's invention occurs by way of an individual only insofar as the multiplicity of imitations and repetitions associated with that individual lead a "life of their own" (Latour & Lépinay, 2009, pp. 37–38). In other words, invention is a collective swarm of *imitative* actions occurring at the level of the trait (Tarde, 1903/2009). This swarm of imitation reconfigures the relations that comprise the individual. A society is monadological in that there are no individuals. Focus on the fold rather than the individual allows Tarde to decenter human agency. Thus, freedom is expressed in relationships to varying degrees, but not as an attribute of individuals. Creativity and invention are simply "a particular moment between invention and adaptation" (Latour & Lépinay, 2009, p. 43). Social Darwinism is avoided, in theory, as long as one stays close to the monads and resists the appeal to master plans, teleological designs, and overarching structures.

Because of the nature of a continuous fold, agents in a monadology don't "interact" with each other, "they own one another to begin with" (Latour, 2012, p. 7). Tarde argues that a monadology operates through the verb "to have" rather than "to be." The French verb *avoir* is used to conjugate most verbs, including verbs associated with subjective feelings, like *to be hungry* (J'ai faime). Tarde's focus on the material verb *to have* rather than the existential verb *to be* is a direct statement of his anti-Cartesianism. In this case, *to own one another to begin with* suggests various material ways of belonging to each other, but not in terms of set containment (not, for instance, as an element within the enclosure of a set), which tends to entrench inside/outside ways of conceptualizing the verb "to have." For Tarde, there is a relational ethics at stake in the monadology that insists we imagine monads in terms of how they *express* the world (folding, grasping, holding). Every monad contains all the other monads insofar as the world is comprised of an infinite continuum of ongoing expression. In other words, a monadology is a vast undulating flow of affective expressivity, contracting and expanding into the infinite past and future. His emphasis on the material verb *to have* allows us to better grasp the temporality of becoming (rather than being), and the contiguity of the past with the future:

> All philosophy hitherto has been based on the verb *Be* [être], the definition of which was the philosopher's stone, which all sought to discover. We may affirm that, if it had been based on the verb to *Have* [avoir], many sterile debates and fruitless intellectual exertions would have been avoided. From this principle, I am, all the subtlety in the world has not made it possible to deduce any existence other than my own: hence the negation of external reality. If however, the postulate *I have* [J'ai] is posited, as the fundamental fact, both that which *has* [eu] and that which *is had* [ayant] are given inseparably at once.
>
> *(Tarde, 1895/2012, p. 52, original emphasis)*

The verb "avoir" captures the *immanence* of life, because it stresses the material coupling and contiguous haptic *encountering* of everyday life (to undergo hunger). In moving away from the Cartesian verb "to be," which seems to be stuck in the dichotomy of being and non-being, monadology operates in and through the materiality of life. Predicates that typically are assigned to being (i.e. I am female) become degrees of having (I have some degree of female). Note how this "having" is not possession of an element in a set, because it is a matter of the degree to which monads express each other. My body expresses female to some degree, and female expresses my body to some degree, and these degrees are not equal. Clearly degrees of power are at work here in configuring the topological features of the monadology:

> To have or to possess is to fold, in other words, to convey what one contains "with a certain power." If the Baroque has often been associated with capitalism, it is because the Baroque is linked to a crisis of property, a crisis that appears at once with the growth of new machines in the social field and the discovery of new living beings in the organism.
>
> *(Deleuze, 1993, p. 110)*

Deleuze directs our attention to how the particular monadology of Leibniz was linked to the particular historical time in which it was articulated. Indeed, as I discuss below, the link between capitalism and monadology becomes newly inflected in our digital times. But before turning to this issue, I raise the question here as to how *knowledge* is conceived in a folding monadology. What could it mean for one monad to know another, if each is always already an expression of all the others? And how might the monadology furnish a new answer to this question, different from the usual postmodern claim that we are always already implicated in our representations of others? First, the monadology studies relationships less in terms of representation, and more in terms of haptic contiguity – or contagion. Second, the monadology offers more than the relativism of postmodernism, which focused on the epistemic limitations of an individual's knowledge of the world. Instead, the monadology affirms an ontological relativism, rather than an epistemological relativism. It asserts that diversity and difference are *not* simply evidence of human limitation but are rather essential forces in the world. This is precisely what Deleuze has in mind when he speaks of *difference in itself*. We can see in Tarde one of Deleuze's inspiration:

> To exist is to differ; difference is, in a sense, the truly substantial side of things; it is at once their own most possession and that which they hold most in common. This must be our starting point, and we must refrain from further explaining this principle, since all things come back to it – including identity, which is more usually, but mistakenly, taken as the point of departure. For identity is only the *minimal degree* of difference and hence a kind of difference, and an infinitely rare kind, as rest is only a special case of movement, and the circle only a particular variety of ellipse.
>
> *(Tarde, 1895/2012, p. 40)*

We might still ask what comes of epistemology if knowledge is no longer or not only the more or less accurate *representation* of the world (Maclure, 2013). As part of this ontological turn, Tarde will claim that an increase in knowledge "about" the world entails tracking the material fold that was previously taken to be a cut. Learning is a process of feeling the contiguous links that are woven together to form the fabric of the monadology. Thus, knowledge is based on haptic encountering (touch) rather than representation and image. *To know* is to track the contiguity *between*, to flow through the connecting lines, and feel the collective affect as it contracts into a knot, or expands and undulates across the surface of a swarm. *To know is to become* this material configuration of proliferating folds and crenellations. I will argue in the proceeding sections of this essay that this monadological approach to epistemology is different from previous postmodernisms insofar as it engages with new digital media. These media are, in fact, fully material and are folded into the ever-changing monadology.

The recombinant subject

Deleuze (1988b) will introduce the term "superfold" to describe the ways in which contemporary digital life is monadological (p. 131). In describing the digital variant of monadology that emerges in our postcybernetic world, he brings together the iterative algorithm of computing power with the genetic fold of life and points to the micropolitics of traits and non-human forces:

> Biology had to take a leap into molecular biology, or dispersed life regroup in the genetic code. Dispersed work had to regroup in third-generation machines, cybernetics and information technology. What would be the forces in play, with which the forces within man would then enter into a relation? It would no longer involve raising to infinity or finitude but an unlimited finity, thereby evoking every situation of force in which a finite number of components yields a practically unlimited diversity of combinations. It would be neither the fold or the unfold that would constitute the active mechanism, but something like the Superfold, as borne out by the foldings proper to the chains of the genetic code, and the potential of silicon in third-generation machines … The forces within man enter into a relation with forces from the outside, those of silicon which supersedes carbon, or genetic components which supersede the organism …. In each case we must study the operations of the superfold, of which the "double helix" is the best-known example.
>
> *(Deleuze, 1988b, pp. 131–132, note that surpli could be translated as overfold)*

Just as the fold was used by Deleuze to describe the Baroque subject, the superfold or overfold is used to describe the current "computerized control society" (Galloway, 2014, p. 108). The dividual (rather than the individual) and the superfold are the key tropes of this new era, still monadological, but reconfiguring the relation between the discrete and the continuous. The term *dividual* refers to the traits that flow through the monadology, across what appears to be an individual member of a society, forging a swarming transindividual society. For Deleuze, the subject emerges through the multiple permutations of these traits – "an unlimited diversity of combinations." If the fold conveys a smoothness, the superfold introduces an algorithmic iteration, a transversal crease in the fabric. The superfold is thus the crease or line that produces a repetition within the monadology. If the fold is "the unit of matter, the smallest element of the labyrinth" (Deleuze, 1993, p. 6), then the superfold is the combinatorial repetition of that fold, the twist or crease of the fold, evoked in the example offered by Deleuze of the double helix. While the Baroque subject was pleated into matter, contemporary dividuals are recombinant subjects, assembling always in relation to the bioinformatic ecosystem. This recombinant subject is dispersed across "an unlimited finity" of molecular repetitions. The superfold will be a fractal fold, an iterative craggy fold, a transversal crease where the flow splits apart but remains attached. The superfold points to a new way of incorporating the discrete into the continuous, and suggests that the act of calculating occurs at the most minute scales.

This fractal monadology seems to offer a philosophical foundation for developing new research methods suitable to these digital times. Galloway (2014) will link this to the current digital dispersion of observation and document, where a multiplicity of points of view proliferate and flood the world, where pixels are reassembled at micro scales, and millions of data points are mined every nano-second. The fractal iteration repeats the fold – where one might think of the fold as the *analog* event – and generates a self-similar somewhat dilated and distorted superfold. It is this iterative repetition of the fold that introduces the discrete into the world. And it is through the discrete that the world begins to count and calibrate. Calculation and numeracy are thus *not* unique human capacities. Instead, calculation is a tendency or potentiality of all matter, be it human or non-human. According to this philosophy, geometry would become a more material mingling of *geo* and *metric,* rather than an idealization or abstraction. For Kirby (2011), too much of socio-cultural discourse theory forecloses this possibility by defining geometry against geology, language against matter, number against object, where the former always codes the latter. Instead, she urges us to consider a *nonhuman mathesis* that disallows any exemption from measure. Of course, there are huge dangers in pursuing this theoretical path! It's all too easy to imagine dystopic scenarios where we pay tribute to a calculating universe with a built-in agenda. We know all too well the horrors of quantitative methods enthrall to cybernetic dreams of calculated publics and digital labor.

But Deleuze helps us see how the control society is not simply imposed on us from without – we come to realize that computation is "within" us (in monadological terms) and that computation is itself a mediated process open to radical reconfiguring. We are computational everywhere, from RNA recombination to markets to digesting stomachs to degrees of affect. These computations are not, however, essentializing, since they are as transient as any other mode of expression. Through this fractal monadology, we begin to grasp the flow and the quanta as mutually entailed. When we consider the digital proliferation of life in our digitized collectives, we can see how a philosophy of immanence is addressed through a fractal monadology. Rather than banish the digital as the *other* to the continuous, or confine their relationship to a dialectic resolution, we can track the mutual-engendering micro-calculations of the flesh with the generative flow of continuous monist matter. Thus, we are tasked to imagine new kinds of quantitative research methods, aside from the usual segmenting of the continuous, where quantity is always assumed to be that which encodes matter. Our research methods must begin to imagine how discrete quantity is somehow immanent within continuous matter. Might it be the case that computation, calculation, and the digital more generally, écho previous modes of communing with matter, as we find in Tarde and Leibniz?

How might such a claim be taken up and developed into a research method? And how would such a claim shift where the incomputable lurks? How might the quantitative merge with this continuous flow, if not in the incremental methods we currently have, manifest in our machine learning models? Indeed, Deleuze and Guattari (1987) will argue for "the importance of statistics, providing it concerns itself with the cutting edges and not only with the "stationary" zone of representations." (p. 219). Such a statistics must pursue the lines of flight, the zig-zag folds that form creases in the monadology and segment the fabric in new transversal directions. But such a statistics must never forget that these creases "exist only by virtue of the flow suffusing them" (p. 218). In other words, we need to attend to the flows that suffuse and modulate the macropolitics of computational practices (de Freitas, 2014; de Freitas et al, 2016). For Deleuze and Guattari (1987), this is precisely why Tarde was interested in the world of detail and the infinitesimal, the miniscule variation of a bureaucratic detail, the minute movements that constituted a sub-representative matter (p. 219). Tarde critiqued Durkheim's sociology of "social norms" for how it failed to address exactly what needed explaining, that being "the similarity of millions of people" (Deleuze & Guattari, 1987, p. 218). Durkheimians claimed that Tarde was more of a psychologist than a sociologist because of this attention to trait and habit. But he was less interested in how one individual copied another, and instead committed to a theory of flow or wave propagation through a swarm or collective, tracking the way that these flows were superimposed and diffracted. Flows, for Tarde, were always comprised of belief and desire, and were quantifiable not as discrete units but as infinitesimal propagation. Attention to flows offers us a way around the individual-structure binary that haunts the Durkheim sociology project, because flows are indifferent to these two constructs. You cannot attribute a flow to an individual or to a structure. In the next section, I discuss how Bruno Latour develops this idea, drawing again from Tarde's monadology.

Calculating matter

Latour (2002, 2010, 2012) lauds Tarde for his radically different vision of what it might mean for a discipline to be quantitative. Latour reads Tarde as offering an alternative to the tradition of parametric statistical modeling that has all too often misrecognized the dynamic nature of society. Indeed, the very public dispute between Tarde and Durkheim regarding what role statistics plays in sociology supports this reading. Tarde considered the social science of his time reflected a particular "patchy statistical apparatus" and that social scientists should think past the methods of statistics to new ways of working with the quantitative (Latour, 2010, p. 152). In Latour's words, the way we work with and conceive the quantitative has huge impact on social theory – "you have the social theory of your statistics" (Latour, 2010, p. 152).

Recall that for Tarde, the sociologist, everything is a society, including human and non-human collectives. Whether a collective of humans, rats, rocks, or shoes, each collective operates as a society. But studies of society have different methods depending on their *material access* to what they study. Hence, there will be important differences between methods, not because of any fundamental ontological difference between human and non-human, but because we humans have different kinds of access to different kinds of society. For Tarde, different methods are demanded because of the scale and relative size of that which is under study. Whenever we study something that operates at a radically different scale to human interaction, we are obliged to use different methods. It's important to note that such a distinction changes with time, as we develop new prosthetic devices and change our understanding of what constitutes human perception. But in any given historical period, the natural sciences concern themselves with activity that is at a non-human scale, and thus they develop the practice of using sampling and case studies and the practice of inductively generalizing to the whole population. This practice treats the generalized law or model as the governing structure that stands outside of the particulars.

Although such methods are adequately suited to the physical sciences, they are not suitable to the social sciences, where we are able to follow the complex relational ontology of particulars without positing an overarching structure. Latour will take up Tarde's point and argue that contemporary social scientists who ape after the natural sciences, in putting such structure to work, fail to grasp the virtue of their privileged access to that which they study. The individual-structure distinction breaks down in the social sciences because the social sciences involve humans who study human society (rather than the society of bees or swamps).[1] For Latour the implications of this are monumental, as it offers a "completely different way of calibrating what should be expected from any science of any society" (Latour, 2010, p. 149).

According to Latour, a new quantitative method, based on Tarde's insights and developments in new media, will resist the tendency to construct laws and models that transcend the multiplicity under study. Such methods will not, according to Latour, inductively generate models that transcend the particular or the components of the network. In these digital times, argues Latour, we do not have to abstract from the particulars to the structural law. If in the past, ethnography was said to attend to the particulars of situated individuals through thick description, allowing for an intimacy and proximity that defined qualitative methods, the digital saturation of most lives and contexts brings the quantitative into that proximate fold. If in the past, the quantitative was that model or code that failed to capture the kind of data collected when up close and intimate with a participant, the internet has completely altered the nature of proximity and intimacy. According to Latour, Tarde anticipates this digital turn, arguing that the quantitative operates at the most minute and proximal scales:

> the more we get into the intimacy of the individual, the more discrete quantities we'll find; and if we move away from the individual towards the aggregate we might begin to *lose* quantities, more and more, along the way because we lack the instruments to collect enough of their quantitative evaluations.
>
> *(Latour, 2010, p. 149)*

In other words, the individual is always already a folded quantitative in that the "one" is never given. This qual-quant collective character of our individuality is based on the monadological fractal fold discussed in the previous section. The challenge is then to follow all these dividuals for how they are aggregated, and to resist moving to a static image of society; we must study the way that dividuals come together to form assemblages without producing a rigid structure. Rather than deploy the quantitative as that which produces rigid structure, we need to reclaim the dynamic and unfinished nature of number (and structure), to consider quantity as itself infinitely elastic. When we make the leap to *rigid overarching structures*, we operate only according to a discrete image of quantity (forgetting the fold), and an impoverished understanding of multiplicity. Monadology is precisely what helps us imagine a different mingling of the discrete and the continuous, collapsing them ontologically. The vast amounts of quantitative "evaluations" that are generated every milli-second produce *data waves* that *are* human society; *while the infinitesimal "interval" is never exhausted by discretization*. In related fashion, a quantum flow cannot be attributed to an individual, because by definition it breaks with the conventions of location or attribution. Such flows circulate stochastically across an assemblage, making them quantifiable through probabilistic measures. A flow partakes of an abstract quanta or "degree of deterritorialization" in the line of flight. Flows are always detached desires or beliefs, imitative and varying as they circulate. Deleuze and Guattari (1987) will argue that a statistics of these flows is essential, "providing it concerns itself with the cutting edges and not only with the 'stationary' zone of representations." (p. 219). Deleuze (1988a) will argue for a "qualitative probabilism" (p. 30) which is an enigmatic term that gestures to a pre-Bayesian incrementalism found in the probability of David Hume, as well as the Baroque probabilism of thinkers like the Jesuit Juan Caramuel (1606–1682) who argued that there were no rigid moral rules.

If at our most personal moments, when we feel most authentic, there are minute calculations occurring at all scales, where does the incomputable hide? One can see in Tarde's monadology an attempt to think bio-power long before the term was used, to recognize the way that the body and the neuron and various other traits and tensors join the material flow of capital, indeed *they comprise the flow of capital as much as any abstract currency*. For Tarde, economy and ecology are entwined. It is not that one is a superstructure and the other a material base. There are no principles guiding economics (no invisible hands or ossified social norms), except for those passionate attachments that happen in the proximal moment of encounter. These moments entail calculations of "value," micro-evaluations of more or less, relational adjustments of assemblages, whereby two adjacent molecules adjust to the milieu of their encounter. Calculations are thus never cut off from the social world, performed in some cold objective way upon the material base, *because the material is always already social but also always harboring the potentiality of the quantitative*. This is not the usual socio-cultural approach, not the "cultural capital" approach of Bourdieu, because Tarde draws principally from a philosophy of immanence. He doesn't interpret or code culture in terms of capitalism, but rather sees capitalism as part of the monadology. For Tarde, the capitalist hydra is just one example or one facet of the pulsing monadology – there is no phase of capitalism out of which we emerge. Capital is part of an ontological process of intensification, by which the monadology folds and creases, interlacing what was previously distant, contaminating what was once unified or isolated, through imitation and invention.

As social science researchers begin to study millions of participants who shared their "natural data" on Twitter, X, Instagram, and other social networks, we begin to glimpse the undulations and involutions of this quantum flow. The calculations spread across a network of relations through "imitative rays" or what Latour and Lépinay (2009) will call "contaminations" (p. 9). Desire and belief are contagious, leaping from adjacent cell to adjacent cell, but never through a context or structure, or any "norm" that seems to sit outside of the immediacy and immanence of the encounter. There is no plan guiding contagion, no structural schemas that determine its progress, and so all we can do as social scientists is study its flow. Everything is potentially a number for Tarde, but it is not that we are simply more able to measure what was previously innumerable. Tarde offers a new concept of number – a new way of thinking about the material nature of quantity: "For him, there already exists in the batter, dare we say, a particular type of quantum that has only an *indirect* link to what economists call the quantifiable" (Latour & Lépinay, 2009, p. 17). It is not that the social scientists are doomed because of their mania for calculating, but that they haven't pursued the proliferation of "tensors" that carry "a vast reserve of quantification" (Latour & Lépinay, 2009, p. 17). Only after multiplying the types of quantum does the quantitative fabric of life come alive. The calculating universe is simply operating at scales that most often escape us, while we wander around within this "swarming of assessments" (Latour & Lépinay, 2009, p. 30):

> It is because of this background of "calculable forces" that the addition of calculative devices, of metrological chains, can have such a performative, explicatory capacity, that they can even become forces of production. It is because the monads calculate at all times and in all possible manners that the addition of calculative devices, which are miniscule prostheses, brings about such a prodigious amplification of evaluations.
>
> *(Latour & Lépinay, 2009, p. 40)*

In the next section, I look more carefully at how this prodigious amplification of evaluation shapes the fractal monadology of today.

Big data and dividuation

Latour (2010, 2012) claims that current techniques for digital navigation through social data may provide us with the kind of methods we need to pursue Tarde's vision of social science. Rather than work with a two-tiered model, between the individual and the aggregate, Latour suggests that digital navigation, large data mining methods, and new visualization software may reflect the principles of Tarde's proposal for a different kind of sociology. Rather than relying on small samples from populations and creating a model that is meant to stand outside the data and represent the data, one can study the long-lasting features of social order from *within* the large data set. Latour describes how the internet search itself operates as a network of *continuous* feedback associations. The search *entity* is defined by a network of other entries, and although these form the *attributes* of the entity, they are also changed through the very act of searching. When I search for "monadology," I generate a list of entries that form the attributes of the name "monadology," but since my search intervenes in the network, through ranking algorithms and IP address tagging, the relationship between entity and attribute is continuously reconfigured. In other words, the name "monadology" is folded into the mesh of its own associations. Extending this analysis to generative algorithms further establishes its relevance. Deep learning algorithms and generative models are simply larger computing machines with built-in loops for modifying weights and increasing familiarity with the data (more on this in Chapters 11 and 13):

> 'Specific' and 'general', 'individual' and 'collective', 'actor' and 'system' are not essential realities but provisional terms that depend rather on the ease with which it is possible to navigate through profiles and to envelope them inside their names.
>
> *(Latour, 2012, p. 4)*

The more cumbersome the navigation, suggests Latour, the more likely we are to introduce an external model for representing the complexity. However, introducing external models to describe multiplicities is what we need to resist doing, according to Latour. Similarly, the concept of rigid *structure* (as something qualitatively different from the particulars) is also to be avoided in the study of collective behavior in digital networks, whether it is a structure assumed to be *a priori* or *emergent*:

> The experience of navigating through profiles available on digital platforms is such that when you move from one entity – the substance – to its network – the attributes – you don't go from the particular to the general, but from particular to more particulars.
>
> *(Latour, 2012, p. 8)*

If there are long-lasting or enduring entities (be they identities, institutions, structures, etc.) within this network, it is not because they are *qualitatively* different from the particulars – they don't stand outside as structural models representing the network. An enduring entity like an institution or gender identity – and thus an aspect of social order – endures because it is repeated with variation (like Tarde's imitation and invention), enough times to be counted as the same, but with enough variation and modulation to be "alive." Such agents or entities are never part of a whole, since there is no whole that isn't always made bigger by a network association. In other words, the proliferating associations that are generated through the search also add more – increase in number – the items in the category. The "whole" expands in time as we navigate through the dataset. Profiles are expanding exponentially when searched.

In this new empiricism, there are no norms that transcend the particulars, just meshworks assembling in a new kind of granular quantitative relationship. This quantitative assembling is no longer based on deterministic parametric models that force the quantitative data to conform to pre-given distributions. One might consider this new approach as a kind of grounded theory for quantitative methods, where the stochastic and probabilistic mutations remain operative at the level of the particular. We can see such a suggestion as an attempt to reclaim quantitative methods as part of a philosophy of immanence, and indeed an opening onto the unscripted future of the qual-quant concept. And this is why Latour so appreciates Tarde – not only does the monadology support Latour's image of interaction as an undulating generative network, but also because Tarde shows how "science is in and of the world it studies. It does not hang over the world from the outside" (Latour, 2010, p. 158).

A research method that tracks the huge quantities of traces of internet activity, following the dynamic aggregate as it grows, will produce a map of convolutions and folds. Latour will suggest we follow the "trajectory of individual innovations" as best we can, or in Deleuze's terms, the dividuals that break off and spread through the network. Imitation and innovation are tracked across the mesh of associations, so that an "entity" comes to be known "by adding more and more items to its profile" (Latour, 2012, p. 7). Latour suggests that this proliferation might be an appropriate way to interpret Tarde's concept of "imitative rays" as the mechanisms by which swarms and networks reproduce and repeat (with difference) that which holds them together (p. 15). Thus, the particular algorithms that we use are hugely significant in determining how we engage with digital data sets. It's important to keep in mind that the searches we perform are "a consequence of the type of technology used for navigating inside datasets" (Latour, 2012, p. 4). Studies of emotion using Twitter data offer an interesting and problematic example, as they tap large data sets regarding a topic that was typically studied in the past using interviews and surveys, two qualitative methods that were unlikely to supply insights into the spontaneous and public nature of circulating affect (Golder & Macy, 2011). Using 300 million tweets, and software for analyzing text for lexical bundles, researchers can create *cartograms* or maps in which the original map of a region is distorted and stretched in accordance with the emotional nature of the tweets (Mislove et al., 2010). Although the software is designed to maintain the original borders of states, the area within borders is scaled in proportion to the number of tweets (coded for emotion) that originate there (see Figure 3.1). The researchers track the contagion of emotion across the region during a 12 hour period, using color codes to show how emotions change over time (see http://www.ccs.neu.edu/home/amislove/twittermood/ for video of this data).

Such an approach studies emotions *impersonally*, as something that circulates across populations. The software limitations direct attention to the transmission rather than the content (the lines rather than the nodes). Human bodies become the medium of emotional expression, and are perhaps no more than contractions of this flowing affect across the network. This focus can be seen as part of the turn away from the phenomenological body of lived experience, toward a rhizomatic network of contaminations (Colebrook, 2014). I've pursued digital mapping experiments focused on affect and atmospheres of emotion in Chapters 6–8. But this focus can also be seen as part of the cybernetic fantasy of perfect transmission, an example of "the small cybernetic honeybee engaged in thoughtless, but communicative, actions" (Halpern, 2014, p. 75). In the cybernetic fantasy, affect becomes information. These early approaches to digital computation were intent on displacing materiality with information (Hayles, 1999). They claimed that computational sciences would no longer have to contend with the question of indexicality – cybernetics was the science of form, leaving materialism behind. Indeed, the cartogram above, "pulse of a nation," seems to achieve this translation.

It is precisely this approach to the digital and the quantitative that is being contested in this essay, as I attempt to reclaim an unscripted computation as immanent to matter. My aim is to seek out a counter-history of calculation and computation (in this case, the one found through Leibniz-Tarde-Deleuze-Latour) that mutates this cybernetic fantasy and materializes differently in different media. As Halpern (2014) suggests, Deleuze walks a thin line and "is involved in an ethical act of excavating this possibility and repeating this cybernetic displacement" but in the interest of producing new opportunities for thought rather than for control (p. 58). For Norbert Weiner and Warren McCullough and other early champions of cybernetics, an obsession with "purpose" and prediction fueled the

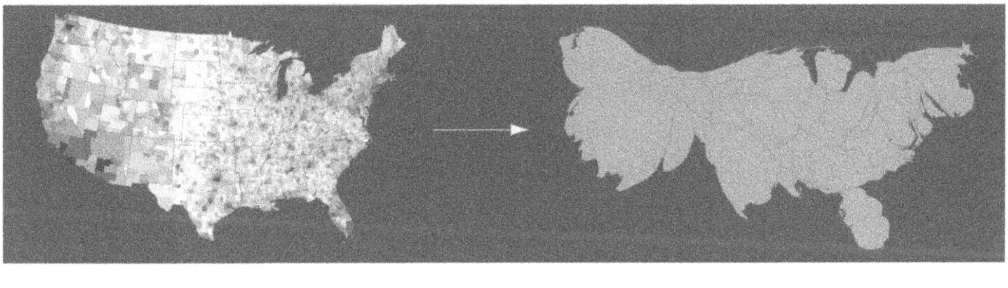

FIGURE 3.1 Cartograms of emotion.

displacement of matter with information (Rosenbueth et al., 1943). Deleuze and Latour critique this *communication* model of interaction, exemplified in cybernetics but found in many language-focused theories of communication, because such models always pay tribute to a logic of exchange, translation, and teleological purpose (de Freitas, 2013). The mistake of cybernetics was to treat communication and computation as equivalent to the act of choosing between discrete units. Although this approach influenced the development of systems theory and various related attempts to think about human society as ecology, and continues to shape so much of our computational culture today, the cybernetic objective was premised on an ideal of instantaneous and unadulterated transmission of information.[2] Instead, Deleuze will reclaim Bergson from Weiner, and show how "the material universe … is the machine assemblage" but without the ideal of communication and its prescriptive futures (Deleuze, 2000). Indeed, Deleuze will propose a calculating matter that doesn't pay tribute to the cybernetic image of communication as a set of choices between two discrete immaterial units. Following Bergson, Whitehead, and other process philosophers, Deleuze will collapse perception and matter, but not as a way of anticipating future signals and predicting future acts of communication.[3]

In the case of the twitter data, it's worth noting that the moving image of the undulating cartogram (available on the website) incorporates a *continuous* visualization of an event using *miniscule repetitive discrete acts*. Cybernetics would see a teleological purpose in this unfolding event, amplified from act to act. Cyberneticians used the term "purpose" to emphasize the goal or target or aim of transmission in any act of communication. When feedback loops from these goals are incorporated into the event (as described above regarding the internet search), they described this as *teleological purpose* (Rosenbueth et al., 1943). It seems crucial to me that we problematize this notion of purpose as we explore alternative imaginaries of computation that refuse to pay lip service to a cybernetic fantasy of control.[4] Another crucial factor in rethinking computation is the issue of the discrete unit. If we treat the word "miniscule" as a term that refers to an objective and stable unit, we adopt the cybernetic tenet that all behavior is commensurable. If instead of the discrete unit we adopt the fold, as discussed in previous sections of this chapter, calculation becomes a plastic topological relation between the folds of the monadology. In other words, the folds of the monadology possess different geometries that are potentially incommensurable – there is no stable unit that can be used to measure each. The curvature of the fold is precisely what allows for different kinds of geometries, where basic notions such as "straight" and basic spatial relationships abide by different axiomatic constraints. Thus a fractal monadology possesses heterogeneous measures across its surface, and there is no constant of proportionality that can be used to measure one by the other. As we look to the future of computational methods in the social sciences, it seems essential that we study this radical incommensurability between measures. I follow Halpern (2014), who suggests we must excavate alternative genealogies in the history of ideas and "reveal these absurd, conflicting, and nondeterministic options for envisioning the future of how we sense and live in data-filled environments" (p. 35).

Concluding comments

The current flourishing of computational distributed networks that link the somatic with the political demands new ways of thinking about how the digital is incorporated into life. Drawing on the sociologist Gabriele Tarde, we find that both Deleuze and Latour advocate for a new kind of monadology that incorporates the digital. I have characterized this monadology as *fractal* because the folds are creased by an iterative imitation or an algorithmic repetition of miniscule quantitative evaluations. If socio-cultural research methods tended to produce *reflections* of that which they studied, the fractal monadology offers an impersonal and anonymous flowing matter, in which "each" monad "reflects nothing, but absorbs all" (Galloway, 2014, p. 141). Monadology is thus a *relational ontology without exchange*, since there are no individuals that might communicate or transfer knowledge. Thus Tarde's focus on "to have" is not about a trait *belonging* to an individual, but is meant to direct our attention to the collective possession of "flow" in itself, across the monadology. The monadology is both the one and the many – a way of rethinking multiplicity and advocating for an "inclusive materialism" and collectivism without exchange (de Freitas & Sinclair, 2014).

Just to clarify, this chapter is not arguing that matter is ultimately digital or binary. This chapter turns to monadology as a way of rethinking the infinite variability of the quantitative. The concept of "the smallest interval" or infinitesimal plays a crucial role in this argument. Monads are ultimately composed of "infinitely small differences," rather than being "the sum of definite and discrete differences" (Tarde, 1895/2012, p. 9). The infinitely small difference or infinitesimal does not differ from the finite only by degree – one *cannot* arrive at the infinitesimal after some countably infinite set of cuts. This makes the infinitesimal "qualitatively different" from the finite (Tarde, 1895/2012, p. 11). One might think of the infinitesimal as force, mobility, duration, the virtual, or potential energy. In the terminology of Deleuze, we can say that the infinitesimal is the *virtual* dimension of the finite. This virtual dimension is that which animates matter – "these tiny beings which we call infinitesimal will be the real *agents*, and these tiny variations which we call infinitesimal will be the real *actions*" (Tarde, 1895/2012, p. 11). As Tarde points out, mathematicians developed the theory of the infinitesimal in order to understand *quantity* and yet they proposed "elements which are not at all quantitative" (Tarde, 1895/2012, p. 11).

I want to close with a few more comments about the ethical issues associated with this new empiricism. Deleuze and Guattari (1987) speak to these ethical issues when they claim that "every politics is simultaneously a *macropolitics* and a *micropolitics*" (italics in original, p. 213). They direct our attention to how the political spreads across all bodies, at diverse scales, tracking traits and tensors across a field of intensity, a non-human field that trembles and quakes with multiplicity. The individuated body – the organic image of the body as organism – is disassembled through Deleuze and Guattari's concept of *becoming imperceptible*. I believe their notion of *micro-politics* and *becoming imperceptible* are crucial as we engage with a fractal monadology in these digital times. Indeed, Tarde will describe the infinitely small infinitesimal as "the imperceptible" (Tarde, 1895/2012, p. 9). A new empiricism of the fractal fold will have to grapple with these indiscernible dimensions of experience.

This molecularization of politics has the potential to radically open up our research to new ways of attending to the biopolitics of life: "It's too easy to be antifascist on the molar level, and not even see the fascist inside you, the fascist you yourself sustain and nourish and cherish with molecules both personal and collective" (Deleuze & Guattari, 1987, p. 215). The links between macro human politics and the biopolitics of flow and fold are yet to be adequately studied. Deleuze and Guattari will argue that forces traverse these various scales and that macro-politics are plugged into micro-politics, through a molecularization of agency:

> Politics on the grand scale can never administer its molar segments without also dealing with the micro injections or infiltrations that work in its favor or present an obstacle to it; indeed, the larger the molar aggregates, the greater the molecularization of the agencies they put into play.
>
> *(Deleuze & Guattari, 1987, p. 204)*

These micro injections and infiltrations (dividuals) are exactly what Tarde was focusing on, in his proposal for a new mixture of quantitative and qualitative social science. My attempt in this chapter has been to elaborate this philosophical tradition of monadology and link it to computational cultures. My hope is that this tradition furnishes us with innovative ideas about how to study and make sense of new qual-quant mixtures of data. And yet, as the history of cybernetics makes clear, attention to the computational dimension of human culture can lead to reductive behaviorism and essentialism. Moreover, there are serious ethical issues evident in digital swarm behavior; we can well imagine how this data is panic-driven, and how it spreads contagiously across various digital networks through re-tweets and broadcasting, resulting in swarm behavior that is often disturbingly destructive. Latour doesn't deny this fact, but nor does he offer counsel for how to proceed.

Notes

1 Presumably this privilege also applies to bees who study bee society.
2 It's important to note that computer science research continues to look at biological systems for how they solve distributed processing and communication problems, with a view to designing human computational networks according to these models. For instance, Navlakha and Bar-Joseph (2014), compare different biological systems for speed of communication and the relative robustness of the network, suggesting that distributed algorithms mimic these structures (see Chapter 11 this volume for more).
3 In an ironic fold, Wiener (1954) referenced the process philosophy of Henri Bergson to advocate for the collapse of matter and perception. Cybernetic theorists will tap the same process philosophers as Deleuze will – Bergson, Whitehead, James. In this we see how historical excavations of the kind achieved by Deleuze are crucial for opening up to different futures.
4 See Jackson (2013) for an alternative approach to purpose.

References

Barad, K. (2007). *Meeting the universe halfway: Quantum physics and the entanglement of matter and meaning*. Durham, NC: Duke University Press.
Bennett, J. (2010). *Vibrant matter*. Durham, NC: Duke University Press.
Braidotti, R. (2013). *The posthuman*. New York: Polity Press.
Colebrook, C. (2014). *Essays on extinction*. Open Humanities Press.
Coole, D., & Frost, S. (2010). *New materialisms: Ontology, agency and politics*. Durham, NC: Duke University Press.
Da Silva (2016). *Fractal thinking*. https://ocean-archive.org/view/1236
de Freitas, E. (2013). What were you thinking? A Deleuzian/Guattarian analysis of communication in the mathematics classroom. *Educational Philosophy and Theory*, 45(3), 287–300.
de Freitas, E. (2014). Diagramming the classroom as topological assemblage. In M. Carlin (Ed.), *For a people-yet-to-come: Deleuze and Guattari, politics and education* (pp. 95–117). New York: Continuum Publishing.

de Freitas, E., & Sinclair, N. (2014). *Mathematics and the body: Material entanglements in the classroom*. Cambridge: Cambridge University Press.
de Freitas, E., Dixon-Roman, E., & Lather, P. (2016). Alternative ontologies of number: Rethinking the quantitative in computational culture. *Cultural Studies – Critical Methodologies*, 16(5), 431–434.
Deleuze, G. (1988a). *Bergsonism*. (Trans. H. Tomlinson & B. Habberjam). New York: Zone Books.
Deleuze, G. (1988b). *Foucault*. (Trans. S. Hand & P. Bové). Minneapolis, MN: University of Minnesota Press.
Deleuze, G. (1993). *The fold: Leibniz and the Baroque*. (Trans. T. Conley). Minneapolis, MN: University of Minnesota Press.
Deleuze, G. (1994). *Difference and repetition*. (Trans. P. Patton). New York: Columbia University Press.
Deleuze, G. (2000). Bergson's conception of difference. In J. Mullarkey (Ed.), *The new Bergson* (pp. 42–64). Manchester, UK: Manchester University Press.
Deleuze, G., & Guattari, F. (1987). *A thousand plateaus*. (Trans. B. Massumi). Minneapolis, MN: University of Minnesota Press.
Eglash, R. (2023). Ethno-biomathematics: A decolonial approach to mathematics at the intersection of human and nonhuman design. M. C. Borba, D. C. Orey (eds.), *Ubiratan D'Ambrosio and Mathematics Education: Advances in Mathematics Education* (pp. 289–303). Switzerland: Springer
Galloway, A. (2014). *Laruelle: Against the digital*. Minneapolis, MN: University of Minnesota Press.
Golder, S. A., & Macy, M. W. (2011). Diurnal and seasonal mood vary with work, sleep and daylength across diverse cultures. *Science*, 333, 1878–1881.
Halpern, O. (2014). *Beautiful data: A history of vision and reason since 1945*. Durham, NC: Duke University Press.
Hayles, C. (1999). *How we became posthuman: Virtual bodies in cybernetics, literature and informatics*. Chicago, IL: University of Chicago Press.
Jackson, A. Y. (2013). Making matter making us: Thinking with Grosz to find freedom in new feminist materialisms. *Gender and Education*, 25(6), 769–775.
Kirby, V. (2011). *Quantum anthropology*. Durham, NC: Duke University Press.
Latour, B. (2002). Gabriele Tarde and the end of the social. In P. Joyce (Ed.), *The social in question. New bearings in history and the social sciences* (pp. 117–132). London: Routledge.
Latour, B. (2010). Tarde's idea of quantification. In M. Candea (Ed.), *The social after Gabrielle Tarde: Debates and assessments* (pp. 147–164). New York: Routledge.
Latour, B. (2012). The whole is always bigger than its parts – A digital test of Gabrielle Tarde's monads. *The British Journal of Sociology*, 63(4), 590–615.
Latour, B., & Lépinay, V. A. (2009). *The science of passionate interests: An introduction to Garbiel Tarde's economic anthropology*. Chicago, IL: Prickly Paradigm Press.
Maclure, M. (2013). Research without representation? Language and materiality in post-qualitative methodology. *International Journal of Qualitative Studies in Education*, 26(6), 658–667.
Massumi, B. (1992). *A users guide to capitalism and schizophrenia: Deviations from Deleuze and Guattari*. Cambridge, MA: MIT Press.
Mislove, A., Lehmann, S., Ahn, Y.-Y., Onnela, J.-P. & Rosenquist, J. N. (2010). *Pulse of the nation: U.S. mood throughout the day inferred from twitter*. https://www.ccs.neu.edu/home/amislove/twittermood/.
Navlakha, S., & Bar-Joseph, Z. (2014). Distributed information processing in biological and computational systems. *Communications of the Association of Computing Machinery*, 58(1), 94–102.
Portanova, S. (2009). The 'minor' arithmetic for rhythm: Imagining digital technologies for dance. In L. Cull (Ed.), *Deleuze and performance* (pp. 240–260). Edinburgh: Edinburgh University Press.
Rosenbueth, R., Wiener, N., & Bigelow, J. (1943). Behavior, purpose and teleology. *Philosophy of Science*, 10, 18–24.
Sorensen, E. (2011). *The materiality of learning: Technology and knowledge in educational practice*. Cambridge, UK: Cambridge University Press.

Strickland, L. (2014). *Leibniz's monadology: A new translation and guide*. Edinburgh: Edinburgh University Press.
Tarde, G. (1895/2012). *Monadology and sociology*. (Trans. T. Lorenc). Melbourne: Re.Press.
Tarde, G. (1903/2009). *The laws of imitation*. (Trans. E. C. Parsons). Charleston, NC: Bibliobazaar.
Wiener, N. (1954). *The human use of human beings: Cybernetics and society* (pp. 15–27). Boston, MA: Houghton Mifflin. Excerpt from *The human use of human beings* by Norbert Wiener (1950).

4

KAREN BARAD'S QUANTUM ONTOLOGY AND POSTHUMAN ETHICS

Queering relationality

Introduction

In her renowned study of Niels Bohr's work in quantum physics, Barad (2007) attempts to recast the relationship between method and concept, claiming that "Bohr's unique contribution is this: he proposes that we understand concepts to be specific material arrangements of experimental apparatuses" (Barad, 2010, p. 253). This offers insight into the ways in which concepts are themselves apparatus. The concept is thus neither a universal ideal *instantiated* in the material plane nor a social construct *abstracted from* the material plane. For Barad, concepts are not detached codes for sorting and naming activity. Nor are they mere distorted reflections of the world, a perspective that feeds into an "aggrieved postmodern antirealism" (Sheldon, 2016, p. 10). This means that research methods – and any use of experimental apparatus - are constitutive of conceptual content in profound ways. In other words, concepts are working material assemblages rather than pure forms subject only to recognition, imposed on formless and inert matter. Her approach aims to encounter and engage with the *conceptual* on the material plane; in other words, this approach refuses the ontological dualism between matter and meaning – concepts are material, and matter is conceptual.

And yet, as Barad (2007) demonstrates, quantum ontology entails an entirely new conceptual mixture. She tracks the gradients and singularities of this mixture in the actual quantum *experiments* performed by Bohr and others, carefully examining the experimental event for how particular quantum concepts thrive and mutate. In other words, she looks to *experimental practice* to show how scientific research entails a particular metaphysics and a particular ethics. Rather than borrow concepts from physics and use them as codes for describing social activity, *she shows how science does philosophy*. To queer the matter-meaning binary, she gets inside the experiments, unpacking the specific material arrangements to make visible the ever-changing conceptual dimension of matter.

I find her analysis of the history of quantum physics compelling, as she provokes one to reconsider our dis/comforting accounts of the "socio-cultural" construction of facts. Her "agential realism" forces us to find the socio-cultural in matter itself. In this chapter, I take

DOI: 10.4324/9781003538769-4

up critical interpretations of her work – primarily Harman (2016) and Žižek (2012) – to better articulate her quantum ontology. My aim is to show how her work articulates a new kind of empiricism for the social sciences, which involves (1) reclaiming the creative and speculative force of experimentation, as a way of reconfiguring our concept-matter mixture, (2) recentering the philosophical problem as a source of inquiry, and (3) mapping a more-than-human quantum relationality.

Diffractive apparatus

Barad's work presents a model for how we might reclaim an experimental practice in the social sciences, as a means of reconfiguring the concept-matter mixture. She directs our attention to *diffractive apparatus* – that is, to devices for investigating ontological questions about quantum causality, temporality, relationality, and life. A diffractive apparatus is designed to produce evidence about our shared quantum ontology. As such, it must occupy or interfere with a plane of generative philosophical problems – it must function as a "device" that helps us plug into problems such as: how do we live post-quantum causality? How do we mesh sub-atomic and organic temporality? In what ways must we reconceive the very notion of relationality in light of quantum science? Experiments that pursue these questions explicitly, that is, interventions that seek (unapologetically) to investigate our shared quantum ontology, recenter philosophical problems as a source for inquiry in the social sciences. Thus, diffractive apparatuses are different from conventional interventions because of this focus on profound philosophical problems about the quantum relationship between matter and meaning.

Barad takes from Bohr the term "phenomenon" to describe the mutual entailment (or inseparability) of (1) the observed and (2) the agencies of observation. Lenz Taguchi (2010) demonstrates how this inseparability helps education researchers study more accurately the intra-activity of learning. Barad adopts the term "diffraction" from optics, and contrasts it with another optical term, reflection, as two competing ways of studying a phenomenon. Diffraction patterns in optics are traces of intersecting waves of light, showing overlapping disturbances of two or more waves. Mazzei (2014) uses this concept to track the way that different theories collide and interact like waves, creating diffractive patterns and new insights about the phenomenon. Barad borrows from Haraway (1997) the philosophical use of this term diffraction to study the shifting patterns that hold in an indeterminate world:

> Diffraction patterns record the history of interaction, interference, reinforcement, difference. Diffraction is about heterogeneous history, not about originals ... Unlike reflections, diffractions do not displace the same elsewhere, in more or less distorted form, thereby giving rise to industries of metaphysics ... Diffraction is a narrative, graphic, psychological, spiritual, and political technology for making consequential meanings.
>
> *(Haraway, 1997, p. 273)*

Insofar as an experiment involves a diffractive device, the experiment becomes a means of mutating concepts and reassembling the world. Such an experiment has consequential meaning and cannot be described as simply a means to test hypotheses. Experiments are significant when they achieve this kind of ontogenetic reassembling of the world. Experiments may prove or disprove a scientific claim, and surely do quite often, but they also enlist all sorts of material forces and mutate all entangled concepts. They make a difference, a "consequential meaning," as Haraway would say. Experiments are risky, creative events that reassemble the world. To suggest that their role is to simply determine the truth or falseness of a claim is to demote the materiality of the experiment, treating the apparatus as merely that which serves to (in)validate the conceptual content associated with the hypothesis. A closer inspection of the experimental tradition, from Galileo onward, shows how experiments *that make a difference* entail more than simply testing hypotheses.

Thomas Young performed the seminal "two-slit" experiment in 1801 to explore the wave and particle dual nature of light. The apparatus directs light through two slits in one screen, and then a pattern of light is generated on the second screen. A scatter pattern indicates particle behavior, while a series of solid lines indicates wave behavior because such lines occur when the waves of light (with definitive wavelength) intersect and amplify (like waves in water). This solid line pattern is called a diffraction pattern. Modern modifications of the double-slit experiment show that *both light and matter* manifest behavior of both waves and particles, and do so according to the fundamentally probabilistic nature of quantum phenomena (Figure 4.1).

Einstein was critical of the particle-wave duality theory of matter, and in an effort to show how the theory would lead to contradictions, proposed the "which slit" thought experiment. The thought experiment involves altering the apparatus slightly so that the initial light source is measured in such a way that one could detect whether a photon was directed to one slit or another. For Heisenberg, this thought experiment pointed to how our intervention in the apparatus "disturbs" the behavior of the photons, that is, disturbs what would have happened in the absence of such an interference. For Heisenberg, quantum physics shows us the limits of our knowledge, the epistemological limits of what we are able to experience or understand. For Bohr, on the other hand, according to Barad, the thought experiment shows exactly what he claimed – that matter can be either particle or wave, depending on the apparatus with which it is assembled. In other

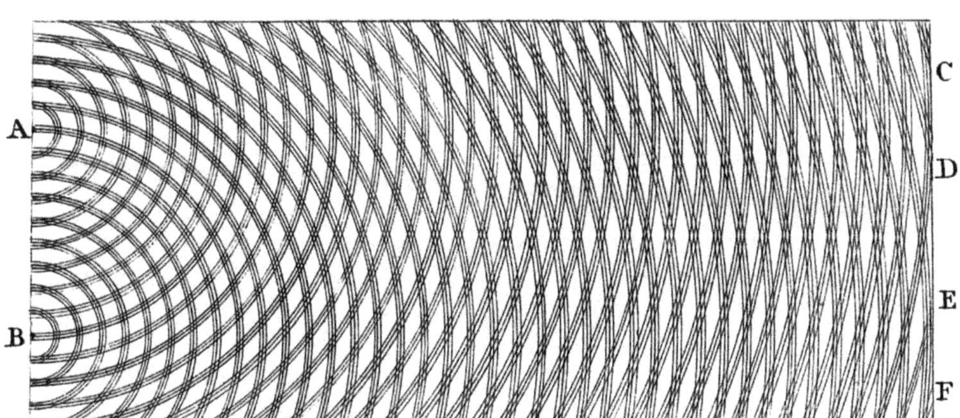

FIGURE 4.1 Thomas Young's sketch of two-slit interference (https://en.wikipedia.org/wiki/Young%27s_interference_experiment).

words, there is not a determinate thing (the unit of matter) that was being observed or capable of being observed at a particular instant. The very notion of observation is put in motion and assembled into the event-nature of the world. Human activity is *implicated* in the ontology of the atomic world, but not simply in terms of an intrusive observation that disturbs what would otherwise have some determinate or unified or fixed behavior (wave *or* particle). Rather, there are no separable isolated entities that can be observed from outside, and thus, "entities" do not have a fixed inherent nature (wave or particle). Duality of wave and particle – and *indeterminacy more generally* – is inherent to matter. This indeterminacy is "in" matter (or simply *is* matter).

Not only does quantum physics undermine the notion of a determinate entity (an entity with determinate spatial boundaries), whereby one body is always spatially entangled with another, but entanglement occurs along the time dimension as well. Barad explains how this works using another famous experiment, the *eraser experiment*, which further develops the *which-slit experiment* by erasing the trace of the photon at the slit *after* it has passed through the slit; that is, by deleting any documented evidence of the fact that we knew which slit it went through. A second (even weirder) eraser experiment deletes this information *after* the scattering pattern on the screen is produced, seemingly further along time's arrow.

If the which-slit information first "caused" the photon to create a particle pattern, and we then erase that information (before or after the photon hits the screen), the pattern alters and becomes a diffraction pattern. Hence, deleting information seems to change, retroactively, the "nature" of the original matter (that nature being particle or wave). This experiment shows how time and not just space is entangled in the apparatus, undermining time's arrow, and troubling the very idea of before and after. *For any given event, there is indeterminacy as to when it occurred or began or ended.* Temporality is itself indeterminate. For Barad, this means "Memory – the pattern of sedimented enfoldings of iterative intra-activity – is written into the fabric of the world. The world "holds" the memory of all traces; or rather, the world is its memory (enfolded materialisation)" (Barad, 2010, p. 261). In discussing these experiments, Barad emphasizes how entanglements entail "irreducible relations of responsibility" (Barad, 2010, p. 265). A diffractive experiment is never a detached observation set in motion merely to confirm an isolated hypothesis. Instead, a new emphasis on experimentation demands a worldly ethics.

More-than-human performativity

Barad (2012c) states very clearly that she is not "applying quantum physics to the social world by drawing analogies between tiny particles and people" for that would be a simplistic misuse of both theory and practice (p. 17). She uses the term "trans/materialities" to describe this complex social entanglement (Barad, 2012c, p. 16):

> Instead of drawing analogies, my method has been to examine the underlying *metaphysical* [emphasis added] assumptions and to understand and elaborate the philosophical structure of the theory. Also, another reason you wouldn't catch me drawing analogies between the two domains is because I question this very idea that there are separate domains of existence.
>
> *(Barad, 2012c, p. 17)*

According to Barad, the focus on radically different scales of being (the quantum or the galactic) does not limit scientific insights into the everyday life of humans. Contrary to the usual way we approach this topic, she emphasizes that there are not two domains (the microscopic and the macroscopic) with two different ontological principles. Thus, she claims that quantum ontology is directly (and not simply analogically) relevant to everyday matters. She uses the term "queer" to describe the behavior of quantum particles, not simply to mark their strangeness, but to tap the meaning of the term in academic discourse, where "queering" has become a methodological way of exploring multiplicity in thought.

Although research methods in the social sciences have begun to turn to the more-than-human in studying intra-activity, Barad suggests this doesn't go far enough and that we need to "find ways to think about the nature of causality, origin, relationality and change without taking these distinctions [human/non-human] to be foundational or holding them in place" (Barad, 2012b, p. 32). She speaks of the "world's performativity" and its "iterative intra-activity," and contrasts her notion of performativity with that of Judith Butler, who characterizes performativity as citational (rather than iterative). The difference between citation and iteration allows Barad to think of a *more-than-human* performativity:

> The inanimate is always being shoved to the side, as if it is too far removed from the human to matter, but that which we call inanimate is still very much bodily and lively. It may seem perverse, unimportant, or meaningless, to attribute memory to an inanimate happening, but that speaks of a failure of imagination that gets stuck at the threshold of one of the most stubborn of all dualisms – the animate/ inanimate dualism – that stops animacy cold in its tracks, leaving rocks, molecules, particles, and other inorganic entities on the other side of death, of the side of those who are denied even the ability to die, despite the fact that particles have finite lifetimes.
>
> *(Barad, 2012c, p. 21)*

But it's not simply the non-human world that thrives and mutates, while the mind or the conceptual is kept pure – even theories are living and breathing reconfigurings of the world: "Theorizing, a form of experimenting, is about being in touch …The world theorizes as well as experiments with itself. Figuring, reconfiguring. Animate and (so-called) inanimate creatures do not merely embody mathematical theories; they do mathematics" (Barad, 2012a, pp. 1–2). Her realism is evident in her claim that the findings of the which-slit and the eraser-slit experiment give "empirical traction" to Derridean theories of trace and texture and performativity (Barad, 2012b, p. 44). By diffractively reading Derrida, that is, *by reading Derrida through the findings of quantum physics*, she argues that we are able to produce a new, *more empirical* deconstruction.[1] Her diffractive method brings deconstruction together with the spooky activity of atoms at the quantum scale:

> Examining the quantum eraser evidence in light of this posthumanist performative understanding of the nature of nature resolves some of the evident paradoxes and gives deconstructionism empirical traction. To put the point differently, this move makes the wager that the radical reverberations of deconstructionism are not merely perverse imaginings of the human mind or of culture but are, in fact, queer happenings of the world.
>
> *(Barad, 2012b, p. 44)*

I take this to be a proposal that "discourse" be treated as a material semiotic and register, a trace or touching, whereby textuality becomes empirical texture:

> While the suggestion that deconstruction might have empirical support may seem blasphemous, to say the least (especially given the common understanding that deconstruction has fully deconstructed empiricism and put it to rest, as it were), Kirby, Schrader, and I are insisting on materialist readings of deconstruction that open up the empirical to reworkings that unmoor it from conventional understandings and do not presume that it can (or even ought to) be put to rest.
>
> *(Barad, 2012b, pp. 45–46)*

Thus, she will go on to describe empirical data as "intelligible speakings of the world" and describe this as a "world worlding itself" (Barad, 2012b, p. 46). Her emphasis on the "speakings" of the world reveals her continued emphasis on discourse, now conceived more materially as trace or register. The indexical register is a kind of sign-making that forms contiguous or material linkages. If language is indexical, it is because there are asignifying particles that make up all expression – language is a kind of haptic/touch relation that *inheres* in the world and perhaps *expresses* the world. Quantum ontology, according to Barad, shows us how language and discourse are haptic encounters. Touch in a post-quantum world reaches beyond the conventional phenomenological framing of hapticity, and suggests a massive intimacy across a spectrum of possible envelopments.

The polymorphous field of concepts

Turning to quantum field theory, Barad uses the term "virtual" to describe quantum particles, such as photons, electrons, Higgs-Boson, etc., but also to describe the plane of trans-touching conceptual dynamism from which a concept is actualized. Any concept actualized in any experiment is haunted by the virtual plane of infinite others that are not actualized. If the concept of circle or cube is produced as young children play with various materials, this concept remains tied to this virtual plane of conceptual dynamism. It's important to note that she does not see the actualized concept as *lacking*, or the act of actualization as a *negation* process. In turning to this virtual dimension of matter, she affirms the *positive* potentiality of the indeterminate. In response to her work, Žižek (2012) offers an alternative interpretation, claiming that quantum indeterminacy marks an inherent negation, lack, or denial that is immanent to the world. Barad instead claims that there is no eternal inert void in which particles exist, because the "void" of indeterminacy becomes a "breathing indeterminacy of non-being" (Barad, 2012a, p. 4). Actualized physical particles intra-act with the virtual particles of the field, and are not independent in the classical sense. In the quote below, we see how she builds on Democritus (the founder of atomism and a chance-inflected universe) to rethink the metaphysics of the quantum world, arguing that the void is no longer "vacuous" but populated, becoming a jubilant field of activity and indeterminacy (Barad, 2012a, p. 4). Accordingly, a polyamorous touching sustains the world through contiguous haptic encounters between the actual and the virtual. Barad suggests that the quantum self-interference of virtual particles is a kind of "self-touching" that sustains the world. She notes that quantum field theory shows how each "individual" always already includes an infinitude of intra-actions with itself *through the infinitude of virtual others*.

All of this highly metaphysical musing helps us understand the ways that quantum ontology is relevant at the human scale. She reclaims and redefines reflexivity in a more-than-human world, using the idea of "self-touching" to describe the fundamental polymorphous perversity of the world:

> Every level of touch, then, is itself touched by all possible others. Hence, *self-touching is an encounter with the infinite alterity of the self. Matter is an enfolding, an involution, it cannot help touching itself, and in this self-touching it comes in contact with the infinite alterity that it is*. Polymorphous perversity raised to an infinite power: talk about a queer intimacy! What is being called into question here is the very nature of the "self," and in terms of not just being but also time. That is, in an important sense, the self is dispersed/diffracted through time and being.
>
> *(Barad, 2012a, pp. 5–6)*

This infinite (and infinitesimal) entanglement informs Barad's approach to ethics. Through this emphasis on "polymorphous perversity," she presents an ethics and a sociality for quantum ontology, and thereby addresses a crucial concern that must be kept at the forefront of any new empiricism. Individuals are indebted to all others through this infinite intra-activity, and this debt, she says, is the condition of all possible giving and receiving. Barad will argue that only through this inhuman virtual indeterminacy can we come to develop an ethics. She follows Derrida and Levinas in proposing an ethics that is prior to all judgment, an ethics that turns on the always already touching of the Other.[2] She suggests that ethics is not about obligation or intentionality, not a calculation of value or utility, but an infinite responsiveness toward that which cannot be sensed – an openness to the "perverse" touch that is occurring beneath or within all the touching of which you are conscious:

> The very nature of matter entails an exposure to the Other. Responsibility is not an obligation that the subject chooses but rather an incarnate relation that precedes the intentionality of consciousness. Responsibility is not a calculation to be performed. It is a relation always already integral to the world's ongoing intra-active becoming and not-becoming. It is an iterative (re)opening up to, an enabling of responsiveness. Not through the realization of some existing possibility, but through the iterative reworking of im/possibility, an on-going rupture.
>
> *(Barad, 2010, p. 265)*

Ethics is thus the capacity to draw on this elemental life that is experienced as inhuman, and to access this inhuman aspect *of human life itself*. We might see this as a kind of eco-ethics that aims to attend to the more-than-human bonds that sustain a diverse material-culture.

It is worth contrasting her approach with that of Žižek (2012), who adopts Barad's ideas about quantum ontology with some important differences. I offer this comparison in order to better appreciate Barad's contribution to the question of how concepts occupy and perform the world. I also think it is essential to offer some detail in this comparison, because Žižek is to some extent a stand-in for a very different theoretical perspective, grounding his work in readings of Lacan and Hegel. This raises once again the issue of whether Barad's work finds an ally in Deleuze or in a very different tradition of Hegelian scholarship represented in the work of Žižek. I explain below how Žižek incorporates particular insights from her work on quantum ontology, but ultimately diverges from what he dismissively calls her "naturalism."[3] He is critical of what seems to be her naïve belief in the natural world – for Žižek, the question regarding where and how concepts are deployed in world making depends on the structure and force of the Lacanian "symbolic order."

Žižek (2012) pursues *analogies* between the quantum and subjectivity, suggesting that the various double-slit experiments point to the power of the symbolic order, which he takes in this case to be the registering of information about the atoms by measuring devices. Just as in quantum physics, where the actual trajectory of a particle "can only be explained if one takes into account all of its possible trajectories within its wave function," the subject of desire, says Žižek, is formed with reference to a vast sea of possible life trajectories: "In both cases, the actualization does not simply abolish the previous panoply of possibilities: what might have happened continues to echo in what actually happens as its virtual background" (Žižek, 2012, p. 920). Žižek takes up Barad's work and suggests that the quantum world is operating much like we operate, occupied by life trajectories and life histories. Similarly, just as the atom seems to know that it is observed, and then changes its behavior (wave or particle), human subjects alter themselves as they imagine themselves seen by others. Furthermore, the time-warp effect that Barad discusses, in relation to the eraser experiment, is shared with the symbolic order, insofar as subjects, often after some intervention or event, will invest retroactively in re-creating their past. Just as an intervention after an experiment can erase past patterns of the atom on the screen, our self-images and life histories can be entirely rescripted after a hugely significant symbolic event.

Such analogies are in fact problematic, as Barad indicates, for they can be naïve projections of physical models from one context to another. These kinds of "subjectivist readings" of quantum phenomena, in which the symbolic is taken to mask a phantom real, are misleading because they assume that the act of measurement distorts what it measures. If it were just a matter of distortion, then one might, in principle, subtract the offending measuring device and get closer to the Real.

The main point for Žižek (2012), however, is that quantum physics studies the *traces* of sub-atomic particles, the "registering" of matter, and is thus always a kind of language or science of language. According to this reading, the two-slit experiments offer evidence of the *immanence* of the symbolic within the real. He sees in the quantum experiments evidence that the world is always already symbolic. Žižek (2012) asks:

> should we risk a step further and claim that there is something which strangely recalls (or points towards) symbolic structures already present in the "physical" reality itself? If we do draw that conclusion, then the entire "spontaneous philosophical ideology" of the gap that separates nature from culture (a form of ideology often clearly discernible in Lacan himself) has to be abandoned.
>
> *(p. 921)*

This claim surely underscores Barad's emphasis on the empirical texturing and performativity of the world. He claims that the symbolic "touches the Real in a totally immanent way" (p. 959). But the symbolic order carries other implications for Žižek, who argues that quantum physics shows that the limitation or lack that characterizes life (and desire) is not limited to the finitude of human beings, but also manifests in the Real. For Žižek, our epistemic obstacle – in not knowing the Real – becomes the limitation of the Real itself. In other words, Barad's fully polyamorous infinite self-touching world becomes for Žižek a world of lack or negation or denial. *Indeterminacy for Žižek is a lack or negation.* This emphasis on lack permeates Žižek's reading of quantum physics, reflecting his Hegelian commitment to the negation of a positive order of being: "What this means, in effect, is that there is no ontology of the Real: the very field of ontology, of the positive order of Being, emerges through the subtraction of the Real" (Žižek, 2012, p. 958). He claims instead that there is a "pre-ontological field" where a universal dialectic produces the inherent lack of the Real. In other words, and quite contrary to Barad, Žižek underwrites the world with a negative dialectic, which undermines whatever plane of immanence and pure difference he was able to glimpse in his discussion of quantum ontology.

Realism and relationality

Barad is a realist about concepts insofar as realists are committed to the existence of concepts as more than mere representations of the "real." As Rouse (2016) states, describing Barad's position: "Conceptual content is not a representation of an object, but a material articulation of a phenomenon, which encompasses both meaning and what is meant" (p. 5). In denying their role as representations, Barad stands apart from psychological approaches that posit concepts as mental constructs in human minds and aligns herself with the contemporary turn to various kinds of realisms and materialisms across the humanities.[4] Harman (2016), however, suggests Barad is far from being a realist because she does not, citing Delanda, "grant reality full autonomy from the human mind" (p. 5). He describes her as a "relationist" rather than a realist, because she invests in the *relationality of concept and object*. In emphasizing the coupling of matter and meaning, Harman sees Barad as someone who is still trapped in a theory that validates "the constant correlation of world and thought" (p. 7). For Harman (2016), following Delanda, a realist is someone who believes that "the world exists independent of minds"; this absolute independence is necessary, according to Harman, if there is to be change and newness, and any chance of a radical "explosive undercurrent" that belongs to the world (p. 22).

What Harman shares with Barad, however, is an affirmation of an agential non-human world; where they differ is in their understanding of relationality and affiliated terms, such as in/dependence. For Harman, any attempt to prioritize "relation" denies the absolute independence of objects and, in so doing, reveals a tacit desire to recenter the human mind in the world. According to this critique, Barad's claim that "only relations exist" is a back-door way to ensure the inclusion of the human in an all too comprehensive encompassing of the world. Despite efforts to decenter the human, Harman argues that a relational ontology is always modeled on human conceptions of relationality.

Barad's quantum ontology, however, presents an entirely non-human relationality that addresses the concern raised by Harman. Moreover, Barad's realism achieves this without banishing concepts as inadequate mental/linguistic representations of the real. In other words, concepts are imbricated in this non-human ontology. Her relational ontology is not a simplistic correlation between world and thought, but rather a philosophy of immanence which attends to the *active* material nature of concepts. Harman is concerned that her approach is "correlationist" insofar as it entails a projection of consciousness onto and into the universe – i.e. she posits a conceptual matter modeled on the human mind (where mind stands in for desire, affect, cognition, and anything else that might be affiliated with human life). He has the same or similar critique for Deleuze. Indeed, one can find sloppy affirmations of a smooth and inclusive "relational ontology" that seem all too self-serving in allowing humans access to everything. All too often, relational ontologies make rather mundane claims about everything being connected or interdependent, without adequately addressing the complex structure of connectivity. I would argue, however, that Barad offers a far more nuanced and complex theory of relationality in a post-quantum world.

Notice how, for Harman, independence between two factors or agents is based on the absence of a *causal* relationship between them. Thus, his critique of relationism is actually based on concerns about conventional Newtonian images of causality and connectivity, which operate across spatial extension and temporal linearity. In quantum mechanics and field theory, such images no longer apply. The quantum leap, discussed below, is a perfect example of how connection is not achieved through continuous extension, at least not in the conventional sense of "continuous." In other words, the space-time continuity of extension that we usually assume necessary for a "relation" to hold, especially one of dependence, is no longer necessary.

In Bohr's model of the atom, electrons are said to *leap* from orbit to orbit (energy level) in such a way that they are nowhere in between orbits during such movement. Such leaps betray the rules of conventional movement: "the electron is initially at one energy level and then it is at another without having been anywhere in between! A quantum leap is a discontinuous movement" (Barad, 2012b, p. 39). Photons of varying frequency, corresponding to the change in energy level of the leap, are released as the electron moves from orbit to orbit. Thus, the photon release is the evidence of movement, and yet the *event* of movement occurred in *no* place. The event happens at a particular time, and yet not in any particular place (for it happens *between* discrete levels, where there is no *continuous* spatial movement, but rather a weird "leap" that is essentially a disappearing act). Moreover, the photon released during the leap is a measure of the energy change (a differential), but under classical physics, this measure can only be achieved after the leap is complete.

It is this weird discontinuous leap that makes her suggest that the non-classical ontology of relationality in quantum physics "might yield a different set of insights about human and non-human worlds" (Barad, 2012c, p. 18). The quantum world introduces a new kind of connectedness that doesn't abide by the old metaphors of relation. Realism has to be redefined in this quantum world. The notion of in/dependence, at the center of Harman's definition of realism (in which the world is said to exist absolutely independently of particular minds, and of thought more generally) also needs to be reconceived.

This queer alteration of relationality leads Barad to use the term "cutting together apart" when discussing processes of individuation.[5] This action of cutting-together-apart is evident in quantum experiments where individual particles, which are seemingly autonomous or independent, are yet entangled with each other and able to "intra-act" instantaneously.[6] Quantum ontology undermines the strict dichotomy between discrete individuation (objects) and continuous connectivity (relations). The queer behavior of quantum leaps is an event that shows us new forms of relationality:

> Quantum leaps are not simply strange because a particle moves discontinuously from one place, here now, to another place, there then, but the fundamental notions of trajectory, movement, space, time, and causality are called into question. And the here and there and now and then are not separate coordinates, but entangled reconfigurings of spacetimemattering.
>
> *(Barad, 2012c, p. 19)*

Barad demands that relationality be reconceived based on these experimental findings. Quantum relationality partakes of a connectivity that breaks with classical physical models, troubling human-scale understanding of space-time events. Rather than affirm wo/man's place in the world, such relationality seems to compose a world indifferent to human action. Then again, if Barad invests in a quantum matter that is "shared" by human and non-human, is she once again performing the anthropocentric move – or at least the anthro-inclusive move – on behalf of her species?

Probability plays a pivotal role in quantum ontology and is indeed at the heart of this new form of relationality – the existence and location of all these leaping particles is bound up with probability measures. An energy orbit is like a sphere of possibility; individuation of an electron is studied through the modal logic of inexact location. Thus, individuation is achieved in terms of *degrees* of existence, using probabilistic metrics and mathematical formulations of conditional relations. Probability is precisely the conceptual content of the quantum leap, and perhaps also the quantitative rumbling that sustains quantum ontology. Notably, however, Barad does not adequately address the absolutely central role of probabilistic reasoning underlying this description of queer quantum behavior. This silence needs attention, as it seems likely that some notion of generative chance, as marshalled within probabilistic forms of activity, may be the key to this new kind of relationality (see Chapter 3). If individuation (of objects or concepts) is a process that engages in the powerful force of probabilistic rendering, then more work is needed in unpacking what Hansen (2015) calls this "probability in the wild."

Concluding comments

I have elaborated the differences between Barad, Harman, and Žižek so that we can better discern the complex web of theory that differentiates them. Their different approaches to relationality, experimentation, and the symbolic reveal important implications for how concept and matter commingle. Barad's quantum ontology, with its emphasis on "cutting together apart," grapples with modal and probabilistic forms of relationality that subvert conventional notions of in/dependence. Harman's concern that relational ontologies always succumb to an anthropocentric image of relationality fails to adequately consider how Barad's ontology "rests" on the queer quantum movement of subatomic matter.

Barad draws our attention to the pivotal role of experimentation in remixing the polymorphous field of concept-method. She looks for experiments with consequential mattering, experiments that pose a philosophical problem. A diffractive apparatus plugs into a plane of generative philosophical problems that make all the present concepts quake and tremble. This approach recenters philosophical problems as a source of inquiry in the social sciences – problems that force us to question the nature of space, time, number, and life. A diffractive experiment investigates our shared quantum ontology by delving into these problems. For Barad, this offers a means to study the empirical traction of Derrida's trace, not as a symbolic lack, but as a kind of symbolic plenitude. Finally, turning to quantum field theory, Barad posits a textured fabric of universal hapticity that threads the actual and the virtual together. Touch becomes the fundamental relation of the world – a quivering quantum tug that holds us together, rather than a classical physical collision encounter. She claims that this quantum touch stretches across the inhuman field of virtual indeterminacy and can furnish an ethics adequate to the world. She leaves us with the challenge of designing and implementing a set of generative experiments that might pursue that aim.

Notes

1 This take-up of deconstruction marks Barad's project as quite different, in lineage, from that of Deleuze or other STS and posthumanist philosophers like Bruno Latour. She does use, however, Latour's terms "matter of fact" and "matter of concern," and I would argue that her interest in an empiricist texturing of the world aligns with Deleuze and Guattari's focus on indexical signs in *Thousand Plateaus*.

2 At points in her work, although perhaps not this one, we can see ways in which her ethics resonates with that of Spinoza. Various scholars have debated the extent to which the basic ethical positions of Levinas and Spinoza are in disagreement.

3 It's important to mention here, in terms of framing the debate and understanding its significance, that in 2012 Žižek publically insulted Barad by dismissing her work as a reflection of her lesbian post-modern tendencies. See E. Geerts and Van der Tuin (2016) for a full elaboration of how Žižek's assessment of Barad reflects an Oedepal dialectical reading (and a politically conservative habit) that runs counter to Geerts and Van de Tuin's diffractive reading.

4 Agential realism (Barad), Speculative realism (Harman), Speculative materialism (Meillessoux), New materialism (Frost).

5 The work of Barad and Deleuze shares many common interests – including the explicit commitment of each to develop new ways of commingling the continuous and the discrete, essentially a new theory of difference. I therefore disagree with S.F. Heine's (2016) characterization of Barad's work as focused on identity and thus entirely at odds with Deleuze's philosophy.

6 Intra-activity is not a form of communication (signal sent and received) but an instantaneous becoming together. The difference is important because the notion of entanglement in physics is not about communication.

References

Barad, K. (2007). *Meeting the universe halfway: Quantum physics and the entanglement of matter and meaning*. Durham, NC: Duke University Press.

Barad, K. (2010). Quantum entanglements and hauntological relations of inheritance: Dis/continuities, spacetime enfoldings, and justice-to-come. *Derrida Today*, 3(2), 240–268.

Barad, K. (2012a). On touching: The inhuman that therefore I am. *Differences: A Journal of Feminist Cultural Studies*, 23(3), 206–223. Revised and republished in *Power of Material/Politics of Materiality* edited by Susanne Witzgall and Kirsten Stakemeier, 153–165.

Barad, K. (2012b). Nature's queer performativity. *Women, Gender and Research*, 1–2, 25–53.

Barad, K. (2012c). Intra-active entanglements: An interview with Karen Barad. *Kvinder, Kon & Forskning*, NR. 1–2, 10–23. Available at koensforskning.soc.ku.dk/kkf/forsidebokse/nyeste/Interview_Karen_Barad.pdf/.

Geerts, E., & Van der Tuin, I. (2016). The feminist future or reading diffractively: How Barad's methodology replaces conflict-based readings of Beauvoir and Irigaray. *Rhizomes: Cultural Studies in Emerging Knowledges*, (30). https://doi.org/10.20415/rhiz/030.e02.

Hansen, M. (2015). *Feedforward: The future of twenty-first century media*. Chicago, IL: University of Chicago Press.

Haraway, D. (1997). *Modest_Witness@Second_Millennium.FemaleMan© Meets_OncoMouse™: Feminism and technoscience*. New York: Routledge.

Harman, G. (2016). Agential and speculative realism: Remarks on Barad's ontology. *Rhizomes: Cultural Studies in Emerging Knowledge*, 30. https://doi.org/10.20415/rhiz/030.e10.

Heine, S.F. (2016). The new materialisms in qualitative inquiry: How compatible are the philosophies of Barad and Deleuze? *Cultural Studies-Critical Methodologies*, 16(2), 132–140.

Lenz Taguchi, H. (2010). *Going beyond the theory/practice divide in early childhood education: Introducing an intra-active pedagogy*. London: Routledge.

Mazzei, L. (2014). Beyond an easy sense: A diffractive analysis. *Qualitative Inquiry*, 20(6), 742–746.

Rouse, J. (2016). The conceptual and ethical normativity of intra-active phenomena. *Rhizomes: Cultural Studies in Emerging Knowledge*, 30. https://doi.org/10.20415/rhiz/030.e01.

Sheldon, R. (2016). Matter and meaning. *Rhizomes: Cultural Studies in Emerging Knowledge*, 30. https://doi.org/10.20415/rhiz/030.e03.

Žižek, S. (2012). *Less than nothing: Hegel and the shadow of dialectical materialism*. New York: Verso.

5

VIDEO DATA AND THE TIME-IMAGE

An-archiving the body

Video research

Video data is currently ubiquitous in social science research. Software protocols for analyzing vast archives of video data are deployed regularly, allowing researchers to annotate, code, and sort images (O'Halloran et al., 2013). Reviews of video research in the Learning Sciences describe how to select an "event" from an extensive video corpus of classroom video, and recommend protocols for how to code these events in terms of student actions, so that one can "interpret events as chunks of time" (Derry et al., 2010, p. 7). These protocols are often applied by researchers without reflection or reference to the extensive philosophical work in film and media studies. The use of video analytic software without adequate attention to how such software is assembles data becomes increasingly problematic as we begin to rely more and more on findings based on this data (Brophy, 2004; Erikson, 2006). This research often treats the video image as movement-image or moving picture, a recording of "raw data," indexical of a given time-space relationship. I myself have used video data exactly in this way. An example is shown in Figure 5.1, which shows a sequence of video stills developed as part of my collaborative work with Francesca Ferrara (de Freitas & Ferrara, 2015).

Such video images are largely used as "realist" snapshots. Many in the field of education research use video to document teaching experiments, design experiments, case

FIGURE 5.1 Students discussing time graphs in mathematics classroom (photo by Francesca Ferrara).

DOI: 10.4324/9781003538769-5

studies, and longitudinal ethnographic studies. More recently, generative software tools such as generative image algorithms like Dall-e are being used to analyze – and in some cases, generate – this kind of data. In this chapter, I situate this kind of research within the history of scientific cinema, and I draw on Deleuze's books on cinema (1986, 1989), as well as his ideas on colour and figure from *Francis Bacon: The logic of sensation* (2005), to propose an alternative way of using video data. My aim is to open up a discussion about how we might engage with video as structured light rather than as realist movement-image. Such an approach follows the artful practice of video, and attends to the materiality of the apparatus. This is part of my ongoing interest in thinking with the technicity of new technologies, and my attempt to look under the hood of research tools. Like the cosmic artisan, only then can we redeploy them as a mode of resistance to simplistic normative applications (see also Chapters 7,8 & 9). One of the central claims of Deleuze in *Cinema 2* is that the time-image reconfigures bodies as expressions of force, as "shocks of force" (p. 139). I argue that such an approach allows us to study video data for how the student body is more than a phenomenological organism with built-in "I can" cognitive and motor capacities. Such a phenomenological perspective is only half the picture. Bodies are also, following a more Deleuzian line of analysis, indeterminate crystalline contractions of intensity. In order to underscore the political significance of both the movement-image within education research and my attempt to shift the analysis, I first discuss how this research inherits particular image conventions from early scientific cinema in which motion efficiency studies served to standardize industrial labour. I then discuss Deleuze's accounts of the history of cinema, emphasizing his turn to the time-image and its potential to disrupt the fixated gaze on sensory-motor bodies. I then present an example of how to study classroom video data as time-image, and explore the implications of such work for social science research.

Scientific cinema

Cartwright (1995) uses the term "scientific cinema" to refer to the early film recordings of moving bodies. Scientific interest in embodiment, according to Cartwright, is indebted to techniques of motion recording and observation developed by experimental physiologists, like Lumière, in France and Germany in the 19th century. Indeed, many of Lumière's contemporaries regarded his invention of the cinematographe as a key contribution to physiology. The Lumière laboratories manufactured film stock and equipment for science – hundreds of films in the Lumière's catalogues cover a vast array of activities or actualities of various kinds of bodily movement. This scientific genre dominated the first decade of film, and can be seen as part of an ongoing interest in capturing and documenting the ephemeral movements of the body. The history of early cinema is thus linked with traditions of scientific experimentation. I emphasize this here, as we explore the ways in which social science is changing in relation to new media. Cartwright (1995) tracks scientific cinema into the 20th century, and argues that studies of the human body in various disciplines need to reckon with this history. She maps "the history of the cinematic techniques that science has used to control, discipline, and construct the human body as a technological network of dynamic systems and forces" (p. 4). This tradition is one that has produced "human life" as a dynamic entity to be tracked and studied through a cinematic lens.

The links between scientific cinema and capitalist modernity are well documented by Doane (2002), who shows how the new technology was used to restructure time and contingency in industrial labour. In the early decades of the 20th century, cinema promised to track the unfolding of time, and to expose the fleeting moment, the "duration of an ephemeral smile or glance" (p. 3). Attempts to capitalize on such knowledge were found within industries of all kinds. For instance, in 1914 Frank B. Gilbreth, a disciple of Taylor (of Taylorism) created "cyclographs" or a "chronocyclegraph" in which expert movements of various labourers were recorded with the aim of creating an archive of efficient motion. In each case, the body's movement was recorded by attaching a small electric light to the limb and using time exposure to generate a representation of the movement as a continuous line in space. In Figure 5.2, a woman is folding a handkerchief. Gilbreth's aim was to study "perfect movement" so as to maximize efficiency of task. These were examples of how scientific cinema was used to create sensory-motor images that exceeded the perceptual capacity of the scientific observer.

According to Doane (2002), the motion efficiency studies served both the standardization of industrial labour and also contributed to an image of time as enacted in bodily movement. Film and photography were considered exceptional technologies for the study of the human body and its unconscious capacities. The indexicality of film – its material coupling to the event – linked the cinematic record to the contingent moment, the any-moment-whatsoever. This link between the technology of the moving image and the capture of contingency was part of the rationalization and abstraction of time that fuelled Taylorism and other capitalist flows. Thus, the capturing of the sensory-motor body in its spontaneous activity was also meant to offer a trace or signature of a certain kind of temporality. Scientific cinema was deployed to capture that temporality and to code the body's movements (Foster, 2004; Sekula, 1986). Although video research in the Learning Sciences

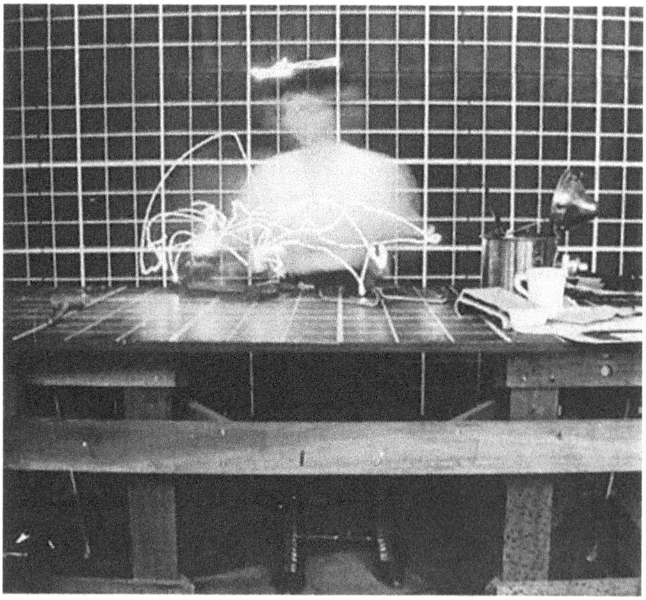

FIGURE 5.2 Motion efficiency study, 1914 (Public Domain image).[2]

continues to track the sensory-motor body and its unconscious ephemeral actions, especially in studies driven by constructivist theories of learning, there are radical differences between it and scientific cinema. Video, for instance, partakes of indexicality in an entirely different way and allows for digital decomposition and reassemblage of images. Of course, montage – as developed by Eisenstein and others – is also a re-assembling of images, more or less experimental in its relation to human perception, but montage is never an obliteration of the image through digital decomposition. I will take up this difference in later sections, when I explore alternative ways of working with such data.

The time-image and the body

According to Deleuze (1989), one can characterize early scientific cinema in terms of the *movement-image*, an image that answers our desire to interpret motor-activity in terms of mechanical cause and effect. I argue in this paper that this same desire fuels contemporary video research and is problematic for how it serves to extend a kind of Taylorism in education research, in some cases actually prescribing particular gestures and bodily movements that facilitate learning (Cook et al., 2013). Such an endeavour seems quite similar to Gibreth's motion efficiency studies and offers substance to Virilio's thesis, as described by Deleuze, that the movement-image was from the beginning linked to the organization of "ordinary fascism" (Deleuze 1989, p. 165).

Deleuze points to a new kind of post-war cinema that breaks with the conventions of the movement-image. This new cinema operates through the *time-image* and connects with bodies in radically different ways. In the time-image, says Deleuze, the body is no longer the obstacle that "separates thought from itself," but precisely the passage or gateway to the unthought – or life itself. The body is no longer the sensory-motor body, the body that lends itself to control, but a body that forces us to think what is concealed from thought. The time-image is that desire to reach the body "before discourses, before words" (p. 172), to let the body speak or express itself without code. Deleuze will announce that "to think is to learn what a non-thinking body is capable of, its capacity, its postures" (p. 189). The time-image *forms an alliance with thought directly* rather than through the cinematic image as mediator.

If in the movement-image, bodies acted and events and cameras followed, the time-image deploys instead pure optical or sound signs, akin, as Deleuze suggests, to how impressionism entailed a conquering of purely optical space. Impressionist painting was said to offer a more direct contact with the real, rather than a conventional resemblance based on human perception. An important part of this shift to the purely optical in cinema relates to how the sensory-motor situation is displaced and becomes "any-space-whatever":

> Any-space-whatever ... is a perfectly singular space, which has merely lost its homogeneity, that is, the principle of its metric relations or the connection of its own parts, so that linkages can be made in an infinite number of ways.
>
> *(Deleuze, 1986, p. 109).*

Concerns for whether there is a subjective or objective camera at work, or to what extent the motor action propels the camera, are thus displaced, as an impersonal optical situation comes to dominate. Deleuze shows how this shift occurs in narrative cinema, but I want to suggest that this shift could be pursued in scientific cinema. What might this mean for video research in the social sciences? I want to push the logic of the time-image into a zone where aberrant involuntary movement, precisely what early cinema aspired to capture, is affirmed as that which "designates time as its direct cause" (Deleuze, 1989, p. 41). In other words, the time-image breaks with the phenomenological image of the body and its desire for presence and directly links to the force and shock of time itself. In social science video practices, this would entail attending to the particular technicity of video, for how it structures light as pure optic image, a means of disassembling the human body, and opening up our analysis to the fundamental difference that reassembles and sustains bodies.

To that end, the notion of the "irrational cut" – the cinematic cut that breaks our tendency to interpret action in terms of singular causal relationships – becomes a helpful concept for rethinking video data. The "irrational cut" creates so much rupture that it cannot belong to what has come before nor what future it seems to prolong. Deleuze takes the concept of the irrational cut from the history of mathematics – the notion of Dedekind cuts – which were an invention by Richard Dedekind in 1858 to define the real number system using open and closed sets. Each real number – each point in the continuum – was to be uniquely identified with a cut separating the real numbers into two sets on either side. If the cut was meant to characterize or determine a rational number, the number belonged to one of the two sets on either side, and the entire number line seemed nicely sorted; but if the cut was meant to determine an irrational number, that number sat between the two sets on either side, excluded, and thus the number belonged to neither set. The irrational number was always on the outside, as Deleuze says, of both sets, which made it strangely unreachable and yet, according to the possibility of getting infinitely close to it, adequately defined by the Dedekind cut. In this sense, the "irrational cut" in the new cinema is truly the place of constitution where a body is virtually produced in the singular space outside of narrative. By thinking with mathematics about cinematic and video editing, following Deleuze, we might be able to crack open the movement-image of conventional video research in the social sciences.

According to Deleuze (1989), a new time-image cinema will work the "dancing seeds" and "luminous dust" of *becoming*, by perturbing perception and producing "the genesis of an unknown body" (p. 201). The early scientific cinema might have glimpsed this radical unknown body, but it is through contemporary cinema that we begin to grasp the prospect of a new kind of *unrecognizable* body. Here, Deleuze refers to Jean-Louis Schefer in *L'homme ordinaire du cinéma* and his thesis that the object of cinema is:

> not to reconstitute a presence of bodies, in perception and action, but to carry out a primordial genesis of bodies in terms of their white, or a black or a grey (or even in terms of colours), in terms of a 'becoming of visible which is not yet a figure, which is not yet an action.'
>
> *(Deleuze, 1989, p. 201)*

This is no longer a body that is the source of its will and action, but instead a Bergsonian body – a zone of indiscernability. This fact points to the undecidability of the body, "the non-choice of the body," the body without organs. Deleuze demands that we follow the flow of intensity that contracts – provisionally – into a body. This is a body that is constantly assembling, dissembling, and reassembling.

In the next section, I explore how researchers might get at this Bergsonian body with video research and new generative software applications. In particular, I am interested in how one might get at a post-phenomenological body and study the forces of affect that flow across classroom events. My aim is to pursue the time-image of the event, to study the virtual image haunting the video data. How might we encounter this crystal image where the genesis of these material relations is protean? What can we do – as researchers – that might allow one to study video data for the crystalline structure that sustains the provisional balance of individuation of a body or an action? How can we provoke or shock our "audience" into thinking the time-image? If we aim to escape from the psychological automata of a movement-image that always seems to serve state-sanctioned control, then we need to be more radical in the way we work with these digital images.

Video experiments and the posthuman body

The method of video analysis discussed here is inspired by the Argentinian video artist Leonardo Solaas (https://solaas.com.ar/). Solaas creates social web applications and generative experiments using various computer algorithms. I used one of his applications, called Doodl, which was an early generative algorithmic drawing tool (circa 2015) that interacts with video images. Doodl is an online drawing "robot" that feeds on digital images. The robot acts like a swarm of "autonomous computational agents that drift endlessly over the surface of the screen, leaving a trace" (Solaas, solaas.ar.com). The algorithm decodes colours in the source image in terms of the force of colour transition. It quantifies the colour intensity in adjacent cells and translates that data into "a force field" that influences the speed and direction of the flowing lines that are drawn. In other words, the "robots" roam randomly (and endlessly) across the image, continuously calculating a relational difference in intensity that is then translated into the speed and direction of the drawn graphic. This is not a representational tool in the strict sense of resemblance, in that the algorithm is generative of slightly altered and unscripted new images due to the randomness of the roaming, as well as reflecting various kinds of parameters that one can change (# of robots, speed, thickness, angle). The result is always a new video that records this randomly roaming incorporation of the video data (Figure 5.3).

The algorithm incorporates the still image – in this case, an image from classroom video – and turns it into a pure event beyond the organic movement-image that is its source. The new video that is generated – as one records the doodl event – goes on indefinitely, as the roaming is endless (as long as power supplies allow), and the doodling becomes harder and harder to discern in the increasingly coloured screen. Thus, we can say that the drawing tool generates a dynamic diagram for this moment in the classroom interaction. When looking at these genetic images, we can begin to reflect on processes of actualization and individuation, slowly building a very layered temporal crystal for each source image that we use. I think of this method as a kind of *colourism*, a method that seems to offer us a way to tap into a new dimension for each "frame" in the video, a dimension of intensity that both cuts away and disinvests in the usual temporal unfolding of the movement-images.

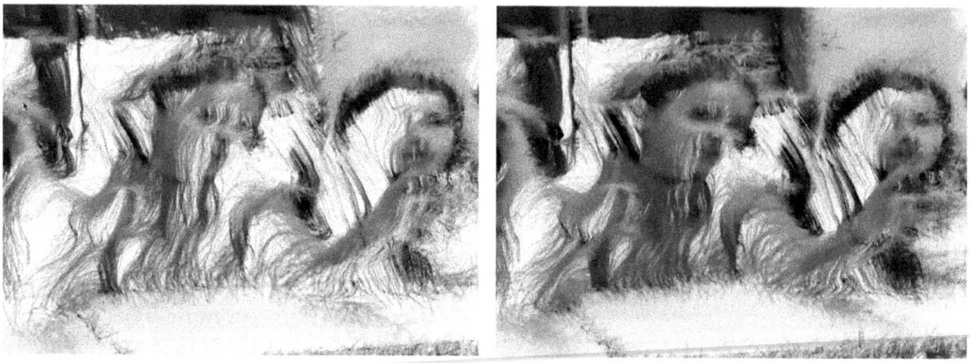

FIGURE 5.3 Two images generated from Doodl analysis of video data (author produced).

When the application begins, streams of flowing, coloured lines emerge, folding and twisting across the initially white screen. Very early on in the application, one notices both the proliferation of undulating lines in a particular colour – where a solid substance slowly emerges – and the formation of border or contour lines that seem to individuate objects, but may simply mark a shadow or a fold in the surfaces as captured in the source image. These fuzzy borders remain indeterminate, no matter how long the application runs, and the smudged effect is radically different from that found in the source image, where we assume the borders around tables, cups, and children are fully articulated.

The processes of individuation that occur as this event infolds operate through the differential of light and its capture in digital quantities coded into pixels. This focus on the intensity and calibration of colour shows the viewer how the gradation of light traverses the various bodies in the classroom. This kind of video analysis lends itself to the study of the materiality of learning assemblages. If light is energy, then we can use this application to study the impressionistic relational ontology of interaction. Not only does this application show the blurring of boundaries between objects, but it shows a *process of actualization* in ways that representational tools do not. The application generates the image using a random walk and a sensitivity to gradations of colour. Unlike a representation of movement, this is a direct presentation of the contingency of embodiment. This application unpacks the stillness of the video still and generates new folds for this now singular moment. This method of video analysis is an infinite process of crystallization, operating through the randomness built into the algorithm, as it "incorporates the environment and forces it to crystallize" (Deleuze, 1989, p. 88).

This painterly rendering of the digital video is an event of infolding, whereby the algorithm actualizes a virtuality haunting the selected image, a virtuality that is linked to the force field of colour/light variation in the video. In *Francis Bacon: The logic of sensation*, Deleuze (2003) discusses how Bacon is interested in the forces that sustain a figure, while wanting to escape figuration as that which is always trapped in narrative. Bacon, like Cezanne, wanted to paint sensation. Deleuze selects Bacon because he dwells on the figure in ways that help us think about sensation in posthumanist ways. As Smith (2003) states in the introduction to the book, Deleuze will return again and again to Bacon's interest in "the figure, the surrounding fields of color, and the contour that separates the two" (p.xii). As a method for analyzing video, Doodl has the same predilections.

As Deleuze says of Francis Bacon's smudge paintings: "create resemblances, but through accidental and non-resembling means" (Deleuze, 2003, p. 80). And indeed these videos and the resulting images evoke the painting of Francis Bacon, operating through a kind of colourism and a disfiguration of the human body, while not completely abandoning the figure.

Doodl attends to relations of tonality. In other words, it maps the modulation of colour intensity in the image. This allows us to study the way different forces are modulated rather than mediated in the particular instant under study. Deleuze (2003) suggests that Bacon's skill at "coloring sensation" is what allows us to get at the posthuman body without organs. He says, "The formula for the colorists is: if you push color to its pure internal relations (hot-cold, expansion-contraction), then you have everything" (p. 112). Colourism aims to show how colour itself is the variable relation or differential relation on which individuation depends. The technique of the colourist is "the production of light and even time through the unlimited activity of color" (p. 112). In pushing colour to function in this onto-generative way, Bacon and other colourists, and perhaps thus Doodl too, force us to encounter the image differently. In watching a doodl video of colouring sensation, the eye is no longer the usual optic device, looking for resemblance, looking for the line, but becomes haptic. In other words, the eye touches the image, and the sense of sight behaves like the sense of touch.

This eerie "scrambling and smudging" of an action-image is a tentative gesture towards the time-image (Deleuze, 2003, p. 127). During each Doodl event, time splits itself into two heterogeneous paths, one that preserves all of the past and another that carries the present into the future. As Deleuze (1989) suggests, "we see in the crystal the perpetual foundation of time, non-chronological time, cronos and not chronos" (p. 81). But we do not always know which is which in this perpetual process of differentiation, and this ambiguity is part of the indiscernibility of the Doodl images. This generative algorithm underscores the many presents that are passing in any given moment, but also directly couples with the past in its entirety and preserves this virtuality in the process of aberrant drawing. In the infolding of colour, we can grasp more than the sequence of events, more than simply the anticipation of action and the coding of gesture as intention, more than the fragment of movement captured in the present moment, and we can begin to grasp as well the ways in which these many presents coexist and the way that all the presents of the past are contracted tightly in the *actual* present.

This method of video analysis demonstrates how it is "we who are internal to time, not the other way round" (Deleuze, 1989, p. 82). Of course, there never is a completed crystal, just as the process of automated random curves can go on forever in Doodl – there is no formula for stopping, save for the reset button. In this sense, we honour the unfinished nature of both thinking and creation. But the crystal image produced by Doodl – relying on the roaming of "robots" that traverse the intensity of the source image – makes visible the differentiation of time in at least two flows. It underscores the present as an infinitely contracted past. Indeed, according to Deleuze (1989), it is this radical contraction that allows for the present to pass on, like the vertex of a Bergsonian cone. Such a form of video analysis explores the empty time of the event, the interstitial time, a different time revealed inside the event, made from the simultaneity of three implicated presents (will happen, is happening, has happened), a threesome that forms the paradox of action. This video method performs the dissolution of the action-image, perhaps similar to what

Deleuze finds in the film-maker Alain Resnais' repetitions, wanderings and immobilizations. In the context of social science research, this video method attends to the temporal depth of all activity, the contraction of time in the singular moment, and the folding of dispersed energy across the field of the classroom.

Rather than a phenomenology that would invoke the "lived body," this method attends to the "profound and almost unlivable power" or "vital power" of sensation (Deleuze, 2003, p. 39). Doubtless, video will prove to be less vitally linked than future poly-sensual recording technologies, and this artisan experiment is already overtaken by newer, more acrobatic generative algorithms, but a video analysis of this sort begins to show how sensation modulates across a classroom, and how activity is dispersed across bodies. Doodl glimpses a block of sensation that is coextensive with perception and representation, generating a new intensive dimension to the screen. This has significant implications for how we consider the non-human dimensions of learning. These kinds of algorithms gesture towards the lines of force or intensity that lie within sensation, underscoring how we are *always becoming nonhuman* just as they show us how we are also actualized as a human figure. I have written about a similar sort of delirium achieved in other generative algorithms that play with our sense of visual representation (de Freitas, 2021, 2022, in press). Such a method of video analysis supports efforts to study the asymmetrical entanglements in classrooms, where assemblages are formed between human and non-human agents (de Freitas, 2012, 2013). As Deleuze suggests, a body is constituted through relationships of force – whether they be chemical, biological, social, or political. Any two forces constitute a body as soon as they enter into a relationship. Thus a living body is the "arbitrary" product of forces of which it is composed (Deleuze, 2006, p. 40). This is why the body is always the fruit of *chance*, says Deleuze, and appears as the most "astonishing" thing, much more astonishing, in fact, than consciousness and spirit.

Two kinds of automata

More work needs to be done on the as yet unknown aspects of the time-image.[1] While the movement-image constituted time in terms of "a successive present in an extrinsic relation of before and after, so that the past is a former present, and the future a present to come" (Deleuze, 1989, p. 271), the time-image is topological, probabilistic, and irrational. In the early scientific cinema, the focus was on capturing movement, which was part of a physiological assumption that one could uncover the fundamental units of movement. In the Learning Sciences, there is an increasing awareness that the scale is sliding, that the micro-gesture hides another micro-gesture, that if one selects another calibration, an infinite fold of time continues to be exposed. Access to finer granularities of time is doubtless linked to particular attributes of the digital image, although not yet clearly operationalized. When we take a video and break it, that is, make it into 24 stills/second or whatever metric we select – and of course, it is that "whatever" which is so important for the automaticity of film – we can no longer invest in the false continuity of the movement-image and we must draw on different energies to make sense of these moving images, especially as we consider using industrial cameras that can record up to 1,000 frames per a second.

Rather than track these frames for how the sensory motor body acts rationally, I am seeking a video research method that will explore how these frames are irrationally linked and potentially *out of order*. Instead of following a linear emplotment, we "turn around" so that time's arrow is twisted into multiple directions. Deleuze (1989) claims that it is "cinema's automatic character" that gives it the capacity to explore the parallel and co-extensive choices that might normally be mutually exclusive. I am advocating for a new video research method which moves beyond the rational linkage of montage to a direct presentation of time, and this entails thought in contact with an unthought – "the unsummable, the inexplicable, the undecidable, the incommensurable" (p. 214). To this, we must add, *the incomputable*. In the time-image, the fissure or interstice has become primary. As Deleuze (1989) suggests, "It is the method of BETWEEN," whether it be between two images, two perceptions, two affections, two actions (p. 180). It is the cut and the calibration "which does away with all the cinema of the One" (p. 180). The objective of this new research method is to make the indiscernible visible, not as presence or will, but in its pure indeterminacy. Although positivist paradigms within the social sciences will judge the suggestion absurd – and for that reason, if no other, we should consider it – I believe that video researchers need to study and learn from cinema and video art. Deleuze points to the director Godard as one who explored the interstices, presenting images that belonged neither to one scene nor another. The insertion of such shots undermines any continuity of internal monologue or plot, making all claims to voice float about unowned in any definitive way. *What can video researchers learn from Godard?*

Deleuze (1989) reminds us that there are at minimum two kinds of automata – the first is the "great spiritual automaton," which pursues the highest exercise of thought, while the second is the "psychological automaton," who is "dispossessed of his own thought" (p. 263). In the first case, cinema confronts automata fundamentally. In the second case, it serves the control society and its reliance on the movement-image. These two kinds of automata are aligned with the two regimes of the image that Deleuze names, the chronic crystal regime (the time-image) and the kinetic organic regime (the movement-image). In the first case, we follow a will to art that allows us to engage with the time-image. This more fundamental encounter breaks with a psychology of conceptual embodiment as well as a phenomenology of the human body as the administrator of all its participation. For the social science researcher, this encounter serves *the will to art*, "aspiring to deploy itself through involuntary movements," but always risking new methods that may destroy that same will (Deleuze, 1989, p. 266). In other words, researchers need to experiment with video to stretch and break the digital image. Rather than or in addition to the slicing of the movement-image into still frames, as early scientific cinema did and as contemporary educational video research continues to do, we need to explore other methods that deploy the digital. Ruppert et al. (2013) suggest that we attend to "the lively, productive and performative qualities of the digital by attending to the specificities of digital devices and how they interact, and sometimes compete, with older devices, ..." (p. 22). They ask that social scientists look to work in the digital humanities and media studies so that we might begin to see how digital technologies can mobilize and materialize new social relations and new research paradigms.

Like so many other recent generative digital tools, the Doodl algorithm was an early example of how to materialize the study of the indeterminate body, a body that is produced through flowing lines and differentials. This body is no longer the centre of wilful action, but the emergent provisional body, a body that exists in potentiality. The human body is studied as a process of individuation, provisionally emergent in relation to the flow of forces across the classroom, itself another body. The student is, nonetheless, still a human body, but one that is always open to new configurations of sensory capacity – indeed, one whose sensory organs are provisionally assembled. The algorithm allows us to pursue the radical implications of cinema's "any-instant-whatever" for, as Deleuze suggests, "when one relates movement to any-moment-whatevers, one must be capable of thinking the production of the new, that is, of the remarkable and the singular, at any one of these moments ..." (Deleuze, 1986, p. 7). The algorithm attends to differences in colour intensity in the video image, incorporating and responding to the subtle gradations encountered in the random walk through the pixel data. These colour gradations partake in the material individuation of bodies in the classroom, slowly vibrating around borders, surfaces and volumes. The undulating lines don't follow a line of human perception, but rather a line of binding intensity.

My hope is that this experimental method of studying video data will act as a catalyst for other experiments in education research. We need to first recognize how our current video practice serves particular kinds of learning theories and particular images of the human body, like other Deleuzian work on research methods in the social sciences has done (see for instance, Coleman & Ringrose, 2013; Maclure et al., 2010). Video research in the Learning Sciences has inherited a set of data conventions from the history of early scientific cinema, a set of conventions that were used in the capitalist over-coding of the human body during 20th century industrialization of labour. We can move beyond these cinematic conventions so that we can begin to map new ontologies of the social and engage with the radical unthought of the time-image. I have continued to integrate historical perspectives on media and the moving image in other video research, where we used the concept of kino-cinema to "bend time" in urban video workshops with young people (see Hickey-Moody et al., 2025). But it also seems urgent that we confront the algorithmic nature of digital devices as we increasingly rely on video data and software protocols. We need to experiment with digital algorithms rather than submit to them, to repurpose them rather than simply submit to the software that is handed down to us. As the software theorist Lev Manovich (2013) asks:

> How does the software we use influence what we express and imagine? Shall we continue to accept the decisions made for us by algorithms if we don't know how they operate? What does it mean to be a citizen of a software society? These and many other important questions are waiting to be analyzed.

My use of *Doodl* is an example of what Luciani Parisi (2013) has called "speculative computation," whereby we pursue "the aesthetic function of algorithms in their quantitative concreteness" (p.xv). The term *speculative computation* supplies a way to think about such broaching experiments, a way of rethinking our relation to the digital. This article offers one breaching experiment that treats video as structured light and gestures towards the multifarious temporal dimensions of the image.

Notes

1 Deleuze, when writing in the 1980s, states of the new digital technologies: "An original will to art has already been defined by us in the change affecting the intelligible content of cinema itself: the substitution of the time-image for the movement-image. So that electronic images will

have to be based on still another will to art, or on as yet unknown aspects of the time-image" (Deleuze 1989, p. 266).

2 *National Museum of American History, Behring Center, Division of Work and Industry Collection.* These three last images found at http://we-make-money-not-art.com/archives/2012/05/the-chronocyclegraph.php.

References

Brophy, J. (Ed.). (2004). *Using video in teacher education*. San Diego, CA: Elsevier, Inc.

Cartwright, L. (1995). *Screening the body: Tracing medicines visual culture*. University of Minnesota Press.

Coleman, R., & Ringrose, J. (2013). *Deleuze and research methodologies*. Edinburgh: Edinburgh University Press.

Cook, S.W., Duffy, R. G., & Fenn, K. M. (2013). Consolidation and transfer of learning after observing hand gesture. *Child Development*, 84(6), 1863–1871.

de Freitas, E. (2012). The classroom as rhizome: New strategies for diagramming knotted interaction. *Qualitative Inquiry*, 18(7), 557–570.

de Freitas, E. (2013). What were you thinking? A Deleuzian/Guattarian analysis of communication in the mathematics classroom. *Educational Philosophy and Theory*, 45(3), 287–300.

de Freitas, E. (2021). Creative abstraction: The role of abduction in machine speculation. *Informatics New Materialism Conference*, Kassel, Germany, March.

de Freitas, E. (2022). Creative abstraction: The powers of the false in algorithmic reason. *Paper Presented at the Annual Meeting of the Society for Social Studies of Science* (4s), Cholula, Mexico. Dec 7–10.

de Freitas, E. (in press). Creativity, deception, delirium. In E. de Freitas, M. Curinga, E. Dixon-Román, & T. Webb (Eds.), *Learning under algorithmic conditions: Contingency, control and creativity*. Minneapolis, MN: University of Minnesota Press. Forthcoming in 2026.

de Freitas, E., & Ferrara, F. (2015). Movement, memory, and mathematics: Henri Bergson and the ontology of learning. *Studies in Philosophy of Education*, 34(6), 565–585.

Deleuze, G. (1986). *Cinema 1, The movement image*. (Trans. H. Tomlinson & B. Habberjam). Minneapolis, MN: University of Minnesota Press.

Deleuze, G. (2006). *Nietzsche & philosophy*. (Trans. H. Tomlinson). New York: Columbia University Press.

Deleuze, G. (1989). *Cinema 2, the time image*. (Trans. H. Tomlinson & R. Galeta). Minneapolis, MN: The University of Minnesota Press.

Deleuze, G. (2003). Francis Bacon: The logic of sensation. (Trans. D. Smith). Minneapolis, MN: The University of Minnesota Press.

Derry, S. J., Pea, R. D., Barron, B., Engle, R. A., Erikson, F., Goldman, R., Hall, R., Koschmann, T., Lemke, J. L., Sherin, M. G., & Sherin, B. L. (2010). Conducting video research in the learning sciences: Guidance on selection, analysis, technology and ethics. *The Journal of the Learning Sciences*, 19(3–53).

Doane, M.-A. (2002). The *emergence of cinematic time: Modernity, contingency, the archive*. Cambridge, MA: Harvard University Press.

Erikson, F. (2006). Definition and analysis of data from videotape: Some research procedures and their rationales. In J. L. Green, G. Camilli, & P. Elmore (Eds.), *Handbook of complementary methods in education research* (pp. 177–192). Mahwah, NJ: Erlbaum.

Foster, H. (2004). An archival impulse. *October*. 110(Fall), 3–22.

Goldman, R., Pea, R., Barron, B., & Derry, S. J. (Eds.). (2007). *Video research in the learning sciences*. Mahwah, NJ: Erlbaum.

Hickey-Moody, A., Pihkala, S., Coombs, G., & Willcox, M. (Eds.). (2025). *New materialist affirmations: Creative research interventions in methods and practice*. Edinburgh University Press.

Maclure, M., Holmes, R., Macrae, C., & Jones, L. (2010). Animating classroom ethnography, overcoming video-fear. *International Journal of Qualitative Studies in Education*, 23(5), 543–556.

Manovich, L. (2013). The algorithms of our lives. *The chronicle review*. Dec. 16, 2013.
O'Halloran, K., (2013). *Multimodal analysis video*. Singapore: Multimodal Software Company. https://espace.curtin.edu.au/handle/20.500.11937/40328
Parisi, L. (2013). *Contagious architecture: Computation, aesthetics and space*. Cambridge, MA: MIT Press.
Ruppert, E., Law, J., & Savage, M. (2013). Re-assembling social science methods: The challenge of digital devices. *Theory, Culture and Society*, 30(4), 22–46.
Sekula, A. (1986). The body and the archive. *October*. (Winter), 3–64.
Smith, D. (2003). Translator's introduction. Deleuze on Bacon: Three conceptual trajectories in The Logic of Sensation. In *Francis Bacon: The logic of Sensation* (pp. vii–xxvii). Minneapolis, MN: University of Minnesota Press.

6

THE BIOPOLITICS OF THE QUANTIFIED SELF

Sensor technologies and worldly sensibility

Introduction

Biosocial research is at the vanguard of a radical and problematic reconfiguring of education research, merging research methodologies from the life sciences and the social sciences (Williamson et al., 2024; Youdell, 2016; Youdell & Lindley, 2019). Such research is increasingly being advocated at national policy levels, responding to scientific developments and paradigm shifts in various disciplines.[1] Many contemporary biosocial interventions, however, are being harnessed to highly conventional and reductionist models of learning, conceived in terms of faster or more effective transit through pre-defined stages. Biotechnologies and "biopedagogies" are being used with children and adults to track and modify attention, engagement, decision-making, emotional states, motion, performance, and creativity (Williamson, 2016). The sensor technologies that are central to this kind of research carry serious ethical implications as they permit new levels of intervention into the bodies, mental states and conduct of individuals and groups (Nafus, 2016). Moreover, behaviour interventions are typically grounded in normative assumptions based on control or correction of bodily phenomena that irritate dominant notions of proper conduct (e.g. "fidgeting," repetitive gestures, noises, agitation, or the turbulence of crowds). Deployments of biotechnologies to track such behaviour are easily, and often rightly, critiqued on ethical and pedagogical grounds (Gillborn, 2016).

It is also important, however, to move beyond the agonistics of critique and towards creative experimentation and the development of new theory. The shift to reconsider biological dimensions is exciting for many reasons, not least in that it allows us to attend to capacities that are often overlooked or poorly understood in conventional educational and social research (Goodman, 2013; Meloni et al., 2016). More generally, the shift in the social sciences towards a rapprochement with biology, ecology, physiology, and other physical sciences opens up opportunities to investigate the material conditions of learning. There are important theoretical developments at play here that could lead to new insights, using the concepts of emergence, thresholds, affective intensities, and creative evolution. So, while I remain deeply suspicious of biosocial frameworks in education,

DOI: 10.4324/9781003538769-6

I am equally committed to thinking with biology and other sciences, drawing from the Latourian image of science discussed in Chapters 2 and 9. My aim here is to take up some of the new insights we have about human bodies and their relationship to their environments, and steal them away from the corporate solutionism that tends to try and own them. I show how scientific and technical knowledge can be redeployed, attending to the aesthetic dimensions of learning environments, tracing affects and sensations that conventional qualitative research methods fail to encounter. This develops an affirmative relationship with some of the sciences. Rose (2013), for instance, pursues

> an 'affirmative relationship' with an emerging 'new and non-reductive biology of human beings and other organisms in their milieu … which can thus be brought into conversation with … the social and human sciences.
>
> *(p. 24)*

This chapter focuses on how a new kind of biosocial research might generate new forms of experimentation that take up scientific knowledge beyond the procedures of normative science. This aim resonates with other projects informed by new materialisms (Fitzgerald & Callard, 2015; Frost, 2016), feminist science and technology studies (Wilson, 2015), and attempts to study nature-culture mixtures (Ingold & Palsson, 2013; Kohn, 2013). In this chapter, I am particularly inspired by the feminist philosopher of science, Elizabeth Wilson (2015), who argues that social theory should become more biochemically literate to reconsider the relational force of pharmaceuticals, metabolism, and the capacity of "biological substances". On the other hand, I am sensitive to the fact that ebullient affirmations of the force of the biochemical without adequate critique, as sometimes found in current work on affect, need to be interrogated (Papoulias & Callard, 2010). Moreover, from my own experience, I am aware of how interdisciplinary projects across the social and biological sciences are often sites where "the biosocial nexus starts to look distinctly bio-centric" (Fitzgerald & Callard, 2015, p. 14).

The primary contribution of this chapter will be in the area of conceptual development around this paradigm shift, with particular focus on how new biotechnology might be used to study a more-than-human worldly sensibility. Specifically, the main aim of this essay is to show how biosensor data might serve in further developing posthuman theories of learning. The term *sensor technology* is used to reference all technologies that produce sensory data about the body or the environment (movement, temperature, air quality, arousal, light, etc.).[2] While avoiding techno-fantasies, either utopian or dystopic, my aim is to trigger imaginings about the potential use of biosensor technologies in education research. I offer an argument as to why social science researchers might reclaim and re-purpose these digital plug-ins, subverting the interests of the control state by showing how such data actually points to the profoundly relational and materially distributed nature of learning. I hope this work contributes to the ongoing process of formulating a robust bioethics adequate to 21st century new empiricisms, by reclaiming biodata that all too often is used to pathologize the individual human body. I argue that digital microsensors produce evidence of a *more-than-human worldly sensibility* and shed light on atmospheres and learning environments rather than individual humans. In essence, this is a politics of refusal – refusing to allow the datafication of the subject – which then re-signifies such data as belonging to the collective, implicating a shared responsibility. This theoretical reframing of biodata is an important step in addressing some of the political concerns associated with the "biosocial subject" (Colebrook, 2014; Protevi, 2013). Although such an approach has its limits, it does allow educators to reframe dis/ability as part of a complex differentiated ecology (de Freitas & Sinclair, 2016).

During the last 50 years, biology has become increasingly computational and data-driven; the new biosocial paradigm is part of this historical development (Stevens, 2013). The use of generative algorithms and large-scale bio data banks has become absolutely central to the biological sciences, as it has for so many other sciences. Computational methods have allowed for all kinds of merging between scientific disciplines, alongside entirely new insights into Anthropocene environments (see Chapter 9 for more on this). Biotechnologies are today more seamlessly integrated into organic processes, and machine learning algorithms are increasingly used in concert with such hardware, suggesting to some that the near future will hold a convergence of the biological, the digital, the sensual, and the social (Platoni, 2015). The historian of science, Freeman Dyson (2007), notes that the "domestication of biotechnology" is transforming our lives and predicts that biology will be to the 21st century what physics was to the 20th. This new Biology 3.0 promises to operate without the biological/digital divide, merging "wet" and "informatic" perspectives into one encompassing biosocial paradigm. With such predictions now almost 20 years old, we are right to be cagey and concerned, and we need to get savvy about social science research methodologies in this new bio-digital-social mixture. How do we study these new strange beings, these digitally "quantified selves" (Lupton, 2014)? And how do so when the term "biosocial" retains problematic associations with past Eugenics movements, and recalls previous misuses of biological sciences (Gravlee, 2009; Meloni et al., 2016). We must keep these historical links in mind as we consider a possible rapprochement between biology and the social sciences.

To that end, my aim is to develop a particular approach to posthuman social science research that shows how digital bio-data belongs to learning environments rather than to individuated organic bodies. This chapter focuses on biosensors that are commonly used to tap the human body for physiological biodata. Wearable sensor technology is now a common way of collecting data from the human body, using new digital microsensors worn on the skin (Coenen et al., 2015). Here I focus on electrodermal bracelets, which are said to offer clinical quality observation and monitoring of temperature, movement, blood-volume pulse, and electrodermal activity (EDA) (Banaee et al., 2013). I discuss in particular one such bracelet produced by Empatica, a company that focuses on "affective computing and data analytics." Empatica designs and sells wearable digital devices for "clinical quality sensing" serving hospitals and research organizations, such as MIT, INTEL, SONY, MICROSOFT, and various universities, wherever there is interest in biometric data.

I work with ideas from Elizabeth Wilson (2015), John Protevi (2013), and Mark Hansen (2015) to offer critique of the applications of such devices but to also argue that such technical intrusions produce evidence of ecological entanglements. This approach recasts biosocial research as the ecological study of complex dynamic systems, where the devices become "atmospheric media" (Hansen, 2015). I argue that electrodermal digital data belong to the charged learning environment and that we must find ways of framing that data in terms of the dynamic fluctuations and entanglements of that environment. This approach troubles the very idea of the organism as the unit of inquiry, and thus differs from related epigenetic research, which continues to focus on the organism insofar as it "responds to the environment" through epigenetic regulation of gene expression. I hope to trouble some of the assumptions of post-cybernetic organicism found in current epigenetic research which treats "the social as signal" and continues to operate within a single causal arrow, albeit granting significant force to the environment.[3] Rather than claim that the environment becomes embodied in the "epigenome" (van Ijzendoorn et al., 2011, p. 309) and rather than focus on trauma and pathology (like most epigenetic research (Meloni, 2016), I read biodata as part of the radical *exteriority of experience*, and as evidence of the inhuman forces at play in any environment. This approach takes up the notion of signal quite differently and studies the inherent *technicity* of matter and life, emphasizing how digital media sense and actuate below the time-scale of the human.

Hansen (2014) argues that digital biosensors plug into a worldly sensibility and suggests that these new digital intrusions are able to access "primordial sensibility" (p. 222) and "enjoy a sensory domain all their own" (p. 234). New mobile media can be studied less as Cyborgian extensions of human faculties and more as registering the "environmentality" of the world (Hansen, 2014, p. 8). On the one hand, digital media produce data that is *about* sensibility (i.e. a sensor records EDA on the skin), while on the other hand, this data *is* sensibility (i.e. a sensor plugs into worldly sensation). Thus, meaning ("about") and being ("is") are digitally coupled in ways that make our particular historical moment distinctive and dangerous. In other words, the current data deluge makes for a singular synchronicity of being and meaning that is apt to be exploited. And yet my own sense is that digital sensors are not mimicking or magnifying human perceptual organs, but instead expanding our understanding of a widely distributed more-than-human block of sensation. Such a perspective entails a "media-driven transformation of human experience itself" and thus a move from an "agent-centered perceptual modality to an environmental sensibility" (Hansen, 2014, p. 8). This paper explores the ways in which sensor data at microscales forces us to imagine life quite differently and to seek the inorganic potentialities and inhuman forces by which a body can "branch out into territories beyond its own self-maintenance" (Colebrook, 2011, p. 26).

Microsensors: data-mining the electric body

The Empatica E4 wristband is designed to record continuous "real-time" data during waking or sleeping hours. It contains a three-axis accelerometer that tracks motion, an infrared thermopile to track temperature, a photoplethysmography sensor (PPG) that measures blood volume pulse (BVP), from which heart rate, heart rate variability (HRV), and other cardiovascular features may be derived. It also contains an EDA sensor "used to measure sympathetic nervous system arousal and to derive features related to stress, engagement, and excitement." Central to my argument is the fact that these biosensors are not operating prosthetically, because they engage with the body in a more distributed and unconscious way, and thus have no correlate to the usual embodied organs, but instead seem to transcend the very notion of organism, while still, paradoxically, mobilizing embodied forces.

The Affective Computing Lab at MIT uses these bracelets in a multitude of projects to study "skin conductance" associated with stressful activity, tracking the variability in how people express stress physiologically. Most of the projects are entirely focused on how such data belongs to an individual human body, and how such data is the expression of affect possessed by that individual human body. In addition, projects affiliated with the lab that are focused on learning assume that such data underscores the cognitive achievements of that individual body. One such project studies children as they play with LEGO blocks (http://affect.media.mit.edu/). Researchers claimed that wrist EDA sensors showed that "children are excited to take on new responsibilities but are then quickly discouraged when they aren't given the resources to succeed." They also claimed that children didn't always recognize their own achievements, based on the EDA data. In other words, the research suggests that skin conductance is a better or more accurate way of determining when children have accomplished something, rather than facial expression or verbal or other visible activity, and that there is a disconnect between these kinds of data, revealing that children *do not know when to value what they have done*.

The aim of the Lego project is uncritically focused on personalized platforms, as the researchers claim that "by using skin conductance sensors, we can help companies better understand the unique perspective of children and build experiences fit for them." The research is explicitly invested in using the EDA data to serve corporations developing curriculum and digital interfaces, as they redesign and "personalize" learning experiences that maximize the individual child's affective engagement, as well as their accurate evaluation of their embodied actions. These aims together reveal how so much of the EDA research inspired by and emerging from the MIT Affective Computing Group is based on a desire to correlate and also control the degree of intensity in any learning experience, and to cultivate self-regulation of affect in children.[4]

At MPath, a company that pursues "Empathic design through rigorous science," EDA data is used to show when people are un/excited, dis/engaged, or stressed. In learning experiments, the data is typically used to show when affect interferes with or supports a goal of some kind. MPath interprets fluctuation in skin conductance as evidence of stress, when, for instance, the EDA graph shows a series of hills and troughs during the experiment. They interpret large singular spikes in EDA data as excitement or severe anxiety, and a trailing off of EDA levels as a sign of disengagement. In Figure 6.1, EDA data from a child is shown, while he uses some building blocks with his mother. The EDA data is said to correlate with two possible scenarios: the first tracking positive excitement, the second negative anxiety. Note the radical difference between these interpretations! *These two interpretations underscore the inherent ambivalence of this data*: either (1) the falling graph marks the boredom of a child as he watches his mother build a block toy, followed by the positive excitement and fluctuation of the EDA when the child works independently. Or (2) the falling graph marks the calmness of the child while he watches his mother build a block toy, followed by the negative stress and fluctuation of the EDA when he anxiously builds independently. The data does not definitively indicate one or the other, although the researchers state that the boy Mason "slouches" and begins chatting about the blocks, losing interest in what his mother is doing during the first minute. Note that this conventional qualitative data is needed to make sense of the digital sensor data.

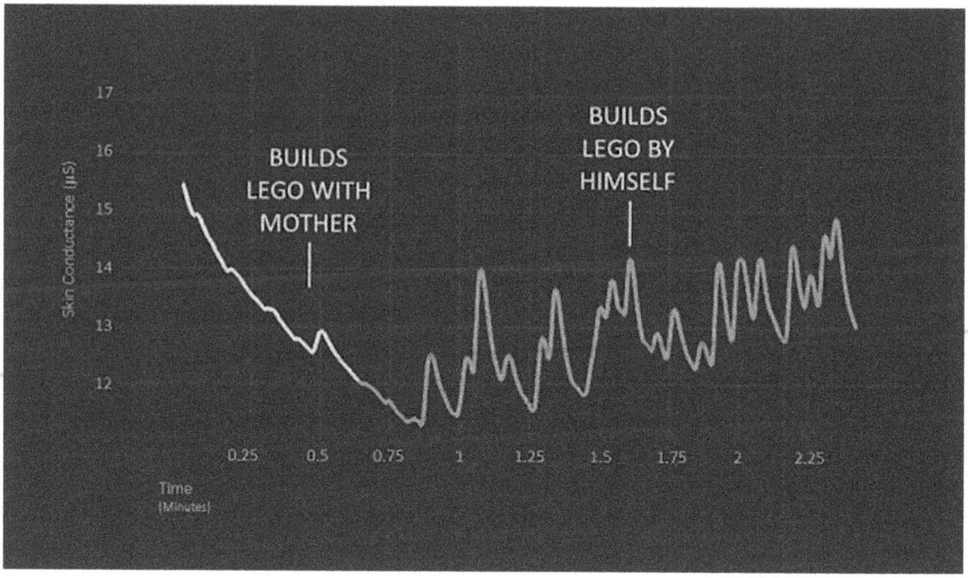

FIGURE 6.1 Measuring engagement.

Source: http://www.buildempathy.com/emototype/.

The main textbook on EDA research was written in 1992 by Boucsein, with a second edition released in 2012. EDA is part of a larger set of data – electrodermal biosignals – which are now the most commonly used data in psychophysiology. Despite the widespread use of such data, electrodermal phenomena are not fully understood. EDA refers to all possible electrical dermal data, including PET and fMRI scans, and thus is at the heart of neuroscience as well as other physiological work today. Electrodermal experiments with humans and animals have a long and checkered history, including horrific public displays of the power of electricity to shock animals into submission (or death), and discredited lie detection devices and other experimental instruments said to correlate electric current with social disposition or competency. Bertucci and Pancaldi (2001) recount episodes in the history of medical electricity, tracking scientific interest in the "electric body" over many centuries. Luigi Galvani's theory of animal electricity, published in 1792, *De viribus electricitatis in motu musculari*, triggered a new line of investigation into electrophysiology (Piccolini & Bresadola, 2013). Even before its publication, in the mid-18th century, when natural philosophy and medical electricity were forging new kinds of empiricism, electric shocks were administered, using Leydon jars, to "penetrate" the patient's body with whatever medical substance was mixed with the water in the jar. I mention this long history of interventions in the electric body to help situate our current interests and media.

With this checkered history in mind, I attempt to reconsider potential uses of the new electrodermal technologies that tap the electric body without violent intervention, in this case "passively" collecting data about the electrophysiology of the body. My argument, however, is not that the EDA data is evidence of an emotion or affect possessed by the human organism, but that skin conductance can be studied as evidence of the radical exteriority of experience, and as evidence of the transitive, relational event-nature of learning. This perspective underscores the need to rethink embodiment, and the need to introduce "a more porous and less self-referential conception of embodiment, a conception that understands the body to be a society of microsensibilities themselves directly and atomically susceptible to technical capture" (Hansen, 2014, p. 193).

The challenge is to explore this work without essentializing these "organic" processes. Classical qualitative researchers would reject this kind of data as irrelevant and dangerous. Rather than dismiss EDA data as irrelevant or insignificant, or serving only to individualize and pathologize the learner, I want to trouble the past antibiologisms of social theory and explore the entanglements of biochemistry, ecology, and learning. My focus on skin conductance is a way of attending to the neurological periphery – the far-flung electrical activity of the body-in-situ – rather than what is assumed to be the centre and administrator of that system, the brain. I am less interested in the central nervous system (the brain, the spinal cord), than in the "distributed network of nerves that innervates the periphery" (Wilson, 2015, p. 5). The electrical charge that innervates the skin is at the periphery of the human body, but it is as central to the activity of rumination, deliberation and comprehension as the brain. This "sympathetic nervous system" is the system that allows the individuated body to belong to a widespread charged environment. The challenge is how to engage with this EDA data without, on the one hand, simply acquiescing to the claim that biodata provides a factual foundation for learning theories, and on the other hand, repeating the doxa of social constructivism and simply dismissing such data.

Wilson (2015) shows how this peripheral nervous system actually dominates the central nervous system in the transmission and distribution of crucial biochemical compounds, such as adrenaline and serotonin. She argues that we must no longer treat "the biological periphery as psychologically inert," nor treat biology itself as inflexible and an obstacle to politics (p. 16). Investing in the speculative potential of matter, she uses recent work in physiology to argue that "biological substance" is as much a "phantasmatic substance" as it is mechanistic (p. 41). She argues that the so-called biological bedrock of the body is robust with both physical and phantastic capacities. The implications for learning theory are risky, as Wilson (2015) calls for the existence of "organic thought" and the "biological unconscious," concepts which have an awkward psychoanalytic history. But these are also concepts that help us problematize the conventional coding of such data in terms of cognitive achievement, or as evidence of an all controlling central nervous system, directing our attention instead to the dispersed nature of affect and thought.

The skin sensors are one way to study this dispersed or distributed phantastic capacity. Indeed the skin occupies the quivering periphery of the "bounded individual" that we take to be the mark of the organism. The EDA skin data is thus perfect for showing how the bounded individual is always being broken down, disassembled, remade, intensified, and charged. Rather than treat synapse and society as disjunctive and antagonistic, one can use the EDA data as a way of witnessing the blended world of the peripheral nervous system. At the juncture of the skin are mixtures of synapse, cilia, sweat, mind, and society, all percolating. Such a reading of the data might be "biological but nonlocalized; chemical but nondeterministic; interior yet worldly" (Wilson, 2015, p. 106).

I suggest that these chemical and electrical actions are a means of modulating worldly sensibility, of ruminating differently with the world, and making the organic periphery tremble. The EDA data points to our biochemical relationality, our bioaffective dispersal, or pulsing electrical belonging. This new empiricism transforms human experience from an agent-centred perceptual modality to an *unequally* distributed "worldly sensibility"

(Hansen, 2015). My use of the "unequally" is important, to remind us that this "distribution of the sensible" is part of socio-economic power relations. We need to develop ways of studying the expanded sensory contact that characterizes this new way of life. In this spirit, I turn to artists who have played with the sensors and the idea of environmental affect. For instance, the urban mapping experiments of Nold (2009) use EDA technology to remap the affective shape of cities all over the world, creating 3-D maps that combined conventional city maps with the graphical peaks and troughs of the EDA data record, produced by volunteers who walked the city wearing sensors.

These kinds of experiments conceive of data differently, not only in terms of biomarkers of personalized affect, but as evidence of an ecological situatedness that disperses the bounded organism. This also involves rethinking the nature of digital biotechnologies, not as "affordances" or "prosthetic" extensions, but more in terms of their actual technicity. I discuss this concept of technicity in more detail in Chapter 13, exploring the ideas of Gilbert Simondon, who helps us better engage with technology – and avoid the two extremes of fear and fetishizing. According to Simondon, a philosopher who influenced Gilles Deleuze tremendously, the virtual and the actual are continuously folded and refolded through a worldly technicity that is not only of our own making. Technology, accordingly, should be analysed as part of more-than-human processes (Galloway & Thacker, 2007; Gane, 2005; Ruppert et al., 2013). In the context of biotechnology and biodata, I also follow Stevens (2013), who suggests that such data "bring the material and the virtual into new relationships" (p. 8). The concept of the virtual is pivotal in clarifying an approach to biosocial data that is not reductive scientism, nor mystical occultism, but is rather based on an alternative way of understanding the material force of the body.

Virtuality, intensity, and the futurity of matter

EDA points to the articulation of bodies as *relations or ratios of speed and rest*. This data is purely differential, insofar as it marks a gradient or rate of change, rather than a definitive quantity that correlates with a particular level of affect. Measures of electrodermal data track *changes* in electrical skin conductance. These changes are linked to the skin's production of sweat, which is itself linked to the sympathetic nervous system. Researchers distinguish between "phasic" data with lots of peaks that seem to mark arousal, and "tonic" data that record gradual changes in engagement (Kim et al., 2013). The fact that there is always this differential element to the electric body helps us theorize a body that is charged, but never static or still – *bodies are related rates of change, each rate itself changing (change of change ...), involving 2nd, 3rd, and nth derivatives*. The peripheral nervous system, extending the body into its frayed periphery, carries this charge in non-stop differentiated flows. *It's as if individuation of a body is a massive related rates problem!* Attending to the electromagnetic field as that which sustains a body, we can begin to study the provisionality of human bodies, and the microtemporality of boundedness. As Bennett (2010) suggests, individuation of bodies "proceeds at a speed or a level below the threshold of human discernment" (p. 58).

Biosensors like the Empatica bracelets detect activity below the perceptual threshold of the human. In this vein, the data testifies to a collective radical exteriority. These EDA devices help us track the provisional ground of embodiment, but also the distributed and differential nature of individuation. In other words, rather than demonize the technology as an extraction device that fails to capture lived experience, we need to find better ways to think about these new kinds of digital plug-ins, different ways of understanding the significance of the EDA data. This involves theorizing the wristbands as a means of plugging into an environmental sensibility, a way of connecting with the machinic dimension of generative activity. The term "machinic" is meant as an alternative to "motor" as in "sensory-motor" which has been used in various phenomenologies that continue to position the brain as the administrator of participation. As Deleuze and Guattari (1987) showed, the machinic better captures the acentred nature of worldly technicity.

Following Protevi's (2013) lead, I argue here that Deleuzian notions of the virtual and of *intensive* individuation can be used to analyse microsensory data. Deleuze elaborates a distinction between the actual and the virtual, as part of his attempt to build a vibrant ontology. By thinking through the role of the virtual in this data, we can begin to think differently about the *capacity* of a body, and this I believe is crucial in developing a bioethics for biosocial research.[5] Deleuze offers a new way of thinking about bodily capacity, less as a fully individuated possibility awaiting realization – achieving its teleological goal of actualization – and more as a live wire or differentiated field of charge. A body's capacity is precisely this terrifying potentiality, this contraction and expansion of forces, this ongoing unexpected worlding. Perception is not the organized synthesis of this sensory surround, but involves another differentiation of already differentiated flows. Protevi (2013) argues that the "virtual web" of "linked rates of change" of neural and other material processes is what characterizes this sensory confound. Crucial to this work is the fact that the relationship between the virtual and the actual is not one of resemblance or determinism, so as to avoid any suggestion of essentializing. As Protevi suggests, there is no relation of *resemblance* between the virtual field and these cascades of differen*t*/*c*iation:

> Thus an individuated perception does not resemble the distributed and differential brain-body-world system, when that is conceived at the level of a virtual web of *linked rates of change* of neural, somatic, and environmental processes.
>
> *(p. 138. My italics)*

Breaking the rule of resemblance is crucial for reclaiming the electrodermal "data" as environmental or ecological rather than only representational of a particular body's properties. But even if we manage to eliminate simplistic notions of resemblance (between sensation and sign), we are still faced with the usual assumption that this microsensory activity determines or causes the more macro bodily activity. This notion of "determination" is at the heart of the dilemma concerning our use of biosocial data in the social sciences, haunting all attempts to bring the biological and the social sciences together. Protevi (2013) seems optimistic, arguing that Deleuze's work resonates with many ideas from the 4EA movement in embodied cognition, drawing primarily from phenomenology and ecological dynamic systems, where 4EA designates embodied, embedded, enacted, extended, and affective.[6] But many of these theories still hold to a notion of the "biologically prior" that determines all biosocial expression. Deleuze's emphasis on a virtual differential potential that is actualized in the machinic phylum undermines this kind of determination, and proposes a different process of determination. For Deleuze (1994), determination is a robust process of differenc/tiation, a term he derives from three key ideas in mathematics (see de Freitas, 2016, for details). These three key aspects of Deleuze's concept of determination (the undetermined, reciprocal determination, the potential) are derived from the calculus and are crucial for how Deleuze moves away from Kant's "conditions" of perception, towards an explanation of how new sensations come into the world.[7]

We can begin to grasp the significance of this theoretical shift by turning now to more general considerations of speed and temporality in biometric research. Consider the microtemporal gap that is said to separate neuronal events from consciousness – a missing fraction of a second between brain activation and awareness. Neuroscientists like Antonio Damasio (2003) suggest that this delay undermines the agency of consciousness – it seems as though conscious discernment is simply that which performs or ratifies what has already been "decided" by the brain. For Damasio, this shows how consciousness is an epiphenomenon, or emergent mental state, a belated effect of its material conditions.

This "fundamental delay" of consciousness in relation to the brain and the body is often cited when researchers work on biodata like that from E4 bracelets. The accelerated speeds of these new biometric devices seem to subvert the slow deliberative time of human consciousness. Instead, Hansen (2015) argues that contemporary digital media of this kind "afford no direct correlation to human perceptual experience whatsoever" (p. 247). He claims that data-gathering from computational microsensors "modulate[s] worldly sensibility directly" without the human subject necessarily recruiting any value from this activity. The humans are on the sidelines – they may indeed be affected, but indirectly because this new kind of digital technicity is, according to Hansen, simply better at plugging into a worldly vibrational sensibility. Humans – with their unwieldy organs – simply are not as good at plugging into that vibrational sensibility.

So rather than chase the simultaneity of sensory solicitation and response, a goal that inevitably serves the "brutal functionalism" (Hansen, 2015, p. 58) of marketing firms, and rather than celebrate the tiny delay as the site of some phenomenological affirmation of our right to slow science, Hansen suggests we study microsensory data for how it plugs into the futurity of matter. Rather than look backwards at the assumed-to-be-complete event of sensation, and our late arrival, look *into* a futural matter that harbours unscripted potential (the virtual). Rather than confine the causal efficacy of sensation to past conditions, take up and analyse the data for how it plugs directly into a robust and ongoing sensibility, a worlding process. The issue is how to avoid or resist the controlling hand of predictive analytics, while still affirming this futural matter. Given how sensor data is already being used to fuel predictive analytics, it seems urgent that we develop this theoretical and practical approach to rethinking the nature of this futurity.

The claim that the living present already includes the tickling agitation of the future is part of the phenomenological tradition, proposed by Edmund Husserl using the term "protention." But Hansen (2015) suggests that protention was largely based on a "ground" of finite possibilities, assembled into memory banks that were then tapped, perhaps much like many machine learning algorithms today must tap vast amounts of training data before they turn to the act of prediction. For Husserl, the human subject accessed this potentiality in the present moment through ratiocination or heightened consciousness. The future was then a sort of projection or a set of expectations based on mental or conscious reflection on this accumulated past. Instead, Hansen (2015), following Alfred Whitehead's philosophy of nature, argues that the future is felt in the present because the future is literally produced (rather than predicted) through the real potentiality of matter. That potentiality is felt (by humans) as *intensity* in the present moment. Intensity becomes the key concept for Hansen, as a complex force (from the future) that animates the present moment. Just as for Deleuze the virtual rumbles perception (and calls forth the actual), for Hansen intensity is the "vibratory character of actuality" and is the feeling by which we sense the future animating the present moment. The intensity of the present moment "simply is the index of the power of this potentiality" (p. 210).

If there is a vibrant futurity animating matter, then we need to rethink forms of *presencing* that are not conventionally phenomenological. The "present moment" is more or less animated or intensified by an undecidable future. This follows Hansen (2015) in suggesting that digital media play a unique role in this situation, as "the scope of the present depends on the degree of precision of technical access" (p. 195). In other words, the degree of that intensity and the specificity of its affect (joy or fear) depends on our access to a worldly technicity that sustains the undecidability of the future. Biosocial studies of "active" learning could then attend more carefully to how biosensors are imbricated in worlding processes. Rather than channels by which signals are relayed, biosensors are somewhat indifferent to human achievement and bodily capacity (Gane, 2005). I suggest that such a perspective brings forth a new politics more adequate to the ubiquitous computing environments in which we now dwell. Reckoning with this digital data deluge and technical milieu requires a new form of resistance. Everything hinges on how well we can live with this "resolutely technical" matter while resisting the ossifying instrumentality that captures and controls it (Hansen, 2015, p. 198).

The biopolitics of precognition

Biopolitics describes how capital saturates everyday life – health, hygiene, sexuality, learning – in order to control populations and to extract new value from new kinds of labour. Foucault (1976) used the term *biopower* to describe the 18th century "explosion of numerous and diverse techniques for achieving the subjugations of bodies and the control of populations" (p. 140). Lazzarato (2002) emphasizes how biopower operates within a complex biopolitical terrain in which control and governance operate along with creative acts of becoming. If earlier versions of biopolitics tended to treat the body as a neutral container shaped entirely by political forces, there is now widespread interest in rethinking the power of the body in contemporary biopolitics (Protevi, 2013). Deleuze and Guattari (1987) offer similar suggestions, offering a way of thinking the political at diverse scales, tracking traits and flows of capital across the molecular and the molar. In their claim that "every politics is simultaneously a *macropolitics* and a *micropolitics*" they shift focus to the molecular and imperceptible level and to what they call "the micro-fascisms" that operate beneath and beside the human subject (*italics in original*, p. 213). This molecularization of politics has the potential to radically open up our research to new ways of attending to the biopolitics of life:

> Politics on the grand scale can never administer its molar segments without also dealing with the micro injections or infiltrations that work in its favor or present an obstacle to it; indeed, the larger the molar aggregates, the greater the molecularization of the agencies they put into play.
>
> *(p. 204)*

These micro-injections and infiltrations break through the skin and reorganize the bounded body according to new forces and new desires. The human body becomes a recombinant subject, engaged in distributed decision-making and network forms of becoming. The individual is divided, repeatedly, replaced by disparate trajectories of "dividuals" – a term that Deleuze and Guattari use to describe the traits that break free from the bounded human body, circulating as part of a biosocial economy. Global capital links the mobile dividuals in a vast network that is highly flexible and transindividual. Thus the individuated human body, upon which theories of labour and capital were built, is disassembled through Deleuze and Guattari's concept of dividuation and *becoming imperceptible*. This approach underscores the way in which capitalism extracts value at all scales and speeds, whether it be from human qualities (race, gender, sexual orientation, etc.) or from the labour of non-human electrical charges as they flow across intra-active currents.

We are all too familiar with the threat of those who capitalize on this kind of "precognitive" biodata. Today's data industries use predictive analytics to target the "operational present" of molecular/neuronal labour as a means of controlling the future. Surely all this granular sensing will lead to nothing else but more control. As Satel and Lilienfeld (2013) argue, it may be that we need to reject this kind of research because it cannot but undermine our commitment to civic, legal, and ethical notions of freedom. Indeed, is it not the case that biosocial research simply deprives us of our ability to shape how are sensibility becomes our experience?

Tentatively, my answer is no. Here are two reasons for not simply resting with suspicion: (1) EDA data is not simply the recording and storing of human bodily experience, but is also a direct engagement or encounter with an electric worldly "sensibility." As argued above, this direct engagement is better studied by attending to the intensity and disparate speeds and dynamics of both individuation and collective coordinated movement; (2) the problem is not so much the technical accessing of biodata but the fact that such data is stolen to serve the control state (Deleuze, 1992). EDA data is here, and we need to deal with it. With these two points in mind, we must not turn our backs on the technology permitting such data capture, but remain vigilant about studying the ways that this technology plugs into a more expansive sensibility.

My approach considers sensibility outside the human-centric notion of perception, as there is no subjective unity capable of hosting such sense-making. If perception is part of learning, it is because the differenc/tiated virtual web of linked speeds is twisted and contracted at particular junctures or events. What is at stake here, then, in EDA research, is a need to think beyond perception studies. Accordingly, Hansen (2015) treats these biosensors as "media" and defines media as that which "operate as instruments that mediate sensibility for experiential achievement" (p. 231). This achievement, however, is indifferent to human embodiment, and might be better studied through some other "climatology" yet to be formulated, tracking the circulation of affect across a charged field (Galloway & Thacker, 2007, p. 157). The aim is to develop a theoretical framework that avoids the two dangerous poles of posturing scientism and mystical obscurantism. In other words, how might we avoid referencing this biodata as the singular cause of individuated ability, while also avoiding the mystical assignment of occult forces at work in matter? If the human subject is a kind of after-image of subpersonal material processes, what is the best way to "make sense" of the virtual or intensive nature of these electrical charges as they flit across the wet surfaces of our bodies? If reductive scientisms turn to biosensors like

E4 to control the future, what kinds of experiments might we design to show how this intensity belongs to the learning environment? What kind of software analytics will help us analyse EDA data as ecological?

Because learning theories are so entangled with theories of perception, I suggest that the biosocial movement might move "from perception-centred accounts of experience to a broader understanding of sensibility as the concrete texture of experience across the board" (Hansen, 2015, p. 48). Indeed, the very notion of "lived experience" – as that fundamental focus of most if not all research on learning – becomes unrecognizable in a world of microtemporal biometric data that circulates and is absorbed at rates well below and above the bandwidth of human consciousness. The crucial thing here is that technology is no longer a surrogate for a human faculty or capacity, but instead operates directly on the sensibility of the "total" environment which precedes and underlies our own corporeal phenomenal experience. The microsensors actually help us study this different/ciated virtual web of linked speeds. In Chapter 13, I turn more explicitly to the philosopher of technology Gilbert Simondon, to further unpack this proposal. The challenge is to take up this data, without making it the biomarker of some essential interiority possessed only or even mostly by the individual, so as to better attend to the trans-individual collective.

Concluding remarks

Quantification of life at scales below and above the human suggests a strange new world of immersive measure (Hayles, 2017; Manovich, 2002; Ruppert et al., 2013). These developments in microsensory digital technology support post-phenomenological studies of embodied learning, and demand new analytic frames that better integrate the qualitative with the quantitative (de Freitas, 2017). Attuning ourselves to the protean quantitative dimension of life is perhaps our best way into the radical rethinking of subjectivity called forth by the digital deluge. Hansen (2015) suggests that "subjectivity" inheres within everything, but that "the massive proliferation of computing in our world" demands that we reckon with how subjectivity inheres in quantitative data as well (p. 12). One wonders whether such an approach resonates with that of Vicky Kirby (2011) and Karen Barad (2017) insofar as these two scholars suggest that the world is investigating itself mathematically, and that "animate and (so-called) inanimate creatures do not merely embody mathematical theories; they do mathematics" (Barad, 2017, pp. 1–2). Even the most seemingly inert, objectified, and datafied event or body is part of a new circuit of sensory links within a qual-quant milieu (de Freitas et al., 2016). In turning to the subjective and social nature of the quantitative, I am not positing a "real" that lurks behind the quantitative, or an authentic subject that fails to be captured by measure. The incomputable is on the move. Instead I argue that sensor data is implicated in (rather than explicating of) a direct (but unequal) sharing of worldly sensibility. This approach aims to rethink the human as an inseparable ingredient of a larger environmental sensory milieu.

In this chapter I offer a different way of analysing EDA data, refusing the various scientisms and psychologisms that serve to centre subjects with privileged access to their biosocial experience. But I realize that the call to decentre the human so as to better encounter a lively world that is indifferent and yet complementary to humans is, well, slightly duplicitous and politically fraught. This is a precarious project – I aim to encounter the more-than-human *so as to imagine* a future for the human. Johnston (2013) claims that these theoretical moves are driven by an ecological politics that has yet to really displace the human and that emphasis on "relational ontologies" is a back-door way of re-inserting the human in the world. Harman (2016) similarly claims that relational ontologies like that of Deleuze and Whitehead are always imagining a relational material world that is saturated with a *human* image of the virtual.

Hansen (2015), however, sees his project as ultimately and unapologetically serving humans. In fact he critiques contemporary theorists working on the posthuman (including the speculative realist Graham Harman and the new materialist Jane Bennett) for arguing a position that is anti-human in their work. Instead, he sees his work as an attempt to "grasp the place of the human within today's media networks" (Hansen, 2015, p. 2). Looking for a "properly elemental conception of the human," we must, suggests Hansen:

> … adopt a radically environmental perspective encompassing human activity as one element among others: such a perspective views human agency just as it does any other type of agency, namely, as internally differentiated, dispersed across various scales and operational divisions, and implicated in and immanent to a total, multi-scalar cosmological situation.
>
> *(p. 2)*

This is not an impersonal objectifying of the world, but a reconceptualizing of subjectivity as "intrinsic to the sensory affordances that inhere in today's networks and media environments" (Hansen, 2015, p. 3). Sensory digital devices like neuro-imaging and electrodermal bracelets and other "smart" sensors produce surplus data that is then tapped by researchers and potentially folded into capitalist circuits of exploitation. However, from my posthuman perspective, we must ask: How can *more than human* research practices serve an expanded notion of justice that might counter the forces of a control society? Rethinking the nature of presence and mediation is crucial as we explore possible answers to that question:

> The idea, then, is that human experience is undergoing change caused by our entanglement with contemporary media environments, and that the directionality of this transformation inverts the long-standing privilege held by humans as the well-nigh unique addressee of media.
>
> *(Hansen, 2015, p. 6)*

This chapter is a call to researchers to pursue counter-deployments of these new media interventions, to reclaim the sensory data stolen from skin, neuron, eye, and electrical impulse, and build a new kind of empiricism that seeks the "production of intensity" while also ensuring an ethical future for labouring human bodies (Hansen, 2015, p. 66). The concept of intensity plays a pivotal role here, because intensity, rather than perception, better expresses the more-than-human forces operating alongside the human: "intensity is precisely what is at stake experientially in twenty-first-century media's direct mediation of worldly sensibility" (p. 103).

As biosocial research maps a new biopolitical terrain, I see a need to shift the focus from theories of agent-centred perceptual capacities to theories of worldly sensibility and "environmentality." Through biosocial research, and related policy, bodes are being reconfigured and reassembled, and the conventional understanding of how a learner learns is changing. Sensor data is precisely the kind of data that all too easily joins the accelerated flows of traits and tensors, as the human organism breaks apart. For this reason, we need to find new ways of studying the material ecology of classrooms and learning environments, refusing to package sensation in terms of biomarkers of dis/ability. New kinds of questions need to be posed by researchers, questions that can help us build more complex models of the charged environment, to avoid being trapped in overly simplistic models of stimulus-response, and to enhance our appreciation for the worldly sensibility that is at stake.

Notes

1 Economic & Social Research Council. (2014). *ESRC framework to enable biosocial research*. http://www.esrc.ac.uk/files/research/framework-to-enable-biosocial-research-pdf/.

2 As Nafus (2016) points out, a broad definition of 'biosensor' helps open up the conversation around subjectivity and self.

3 Epigenetic research has inherited cybernetic notions of receptor and relay, so that the environment now 'signals' and is transduced through bodies (Landeker, 2016).

4 The lead MIT researcher at the lab, Rosalind Picard, founder of the company Affectiva, is interested in how EDA might help Autists, and people suffering from seizures, anticipate and thereby avoid traumatic incidents. For other examples of research in this area, see Choi, Ahmed, & Gutierrez-Osuna (2012), Hernandez et al. (2014), and Sano and Picard (2013).

5 Hansen (2015) – whose work we discuss in more detail below – critiques Deleuze and the concept of the virtual, and emphasizes intensity instead. Indeed, Hansen is keen to differentiate himself from many other theorists. Despite these claims, my reading of Deleuze – his work on both virtuality and intensity – shows the relevance of his work.

6 Protevi cites Anthony Chemero (2009), Alva Noe (2004), and Michael Wheeler (2005). He then adds and complicates this work in embodied cognition by introducing ideas from Gilles Deleuze (in particular the concept of the virtual).

7 Although not adequate room here to develop this idea, there are important ways in which the concept of the virtual offers a twist to conventional theories of emergence. There is also increasingly more work on quantum cognition and quantum sociology, offering alternative ways of conceiving of 'determination' in contrast to the concept of 'emergence' as articulated in dynamic systems theory see, for instance, Alexander Wendt (2015).

References

Banaee, H., Ahmed, M. U., & Loutfi, A. (2013). Data mining for wearable sensors in health monitoring systems: A review of recent trends and challenges. *Sensors*, 13(12), 17472–17500.

Barad, K. (2017). When two hands touch, how close are they? On touching the inhuman that therefore I am. In K. Stakemeier & S. Witzgall (Eds.), *Power of material/politics of materiality* (pp. 153–165). Zurich: Diaphanes. Available at https://www.diaphanes.de/titel/ontouching-the-inhumanthat-therefore-i-am-v1-1-3075 (accessed 29 August 2017). Originally published in *Differences: A Journal of Feminist Cultural Studies*, 23(3), 206–223.

Bennett, J. (2010). *Vibrant matter*. Durham, NC: Duke University Press.

Bertucci, P., & Pancaldi, G. (2001). *Electric bodies: Episodes in the history of medical electricity. Bologna Studies of History of Science*, 9. Bologna, Italy: University of Bologna.

Chemero, A. (2009). *Radical embodied cognitive science*. Cambridge: MIT Press.

Choi, J., Ahmed, B., & Gutierrez-Osuna, R. (2012). Development and evaluation of an ambulatory stress monitor based on wearable sensors. *Information Technology in Biomedicine, IEEE Transactions On*, 16(2), 279–286.

Coenen, T., Coorevits, L., & Lievens, B. (2015). The wearable Living Lab: How wearables could support Living Lab projects. Presented at the *Open Living Lab Days 2015*, Istanbul, Turkey.

Colebrook, C. (2011). *Essays on extinction*. Open Humanities Press. Creative Commons. Available at: www.openhumanitiespress.org/books/titles/sex-after-life.

Colebrook, C. (2014). Socially invaded: The biosocial subject. *Environment and Planning: Society and Space*, 32, 556–570. Available at https://journals.sagepub.com/doi/pdf/10.1068/d3203rev (accessed 30 October 2017).

Damasio, A. (2003). *Looking for Spinoza: Joy, sorrow and the feeling brain*. New York: Harcourt Inc.

de Freitas, E. (2016). Deleuze, ontology and mathematics. *Encyclopedia of Educational Philosophy and Theory*. Available at https://link.springer.com/referenceworkentry/10.1007%2F978-981-287-532-7_374-1.

de Freitas, E. (2017). The temporal fabric of research method: Posthuman social science and the digital data deluge. *Research in Education*. Available at https://journals.sagepub.com/toc/riea/98/1.

de Freitas, E., Dixon-Román, E., & Lather, P. (2016). Alternative ontologies of number: Rethinking the quantitative in computational culture. *Cultural Studies-Critical Methodologies*, 16(4), 1–4.

de Freitas, E., & Sinclair, N. (2016). The cognitive labour of mathematics dis/ability: Neurocognitive approaches to number sense. *International Journal of Education Research*, 79, 220–230.

Deleuze, G. (1992). Postscript on the societies of control. *October*. 59, 3–7. https://doi.org/10.2307/778828.

Deleuze, G. (1994). *Difference and repetition*. (Trans. P. Patton). New York: Columbia University Press.

Deleuze, G., & Guattari, F. (1987). *A thousand plateaus: Capitalism and schizophrenia*. (Trans. B. Massumi). Minneapolis, MI: University of Minneapolis Press.

Dyson, F. (2007). Our biotech future. *New York Review of Books*. July Issue. Available at https://www.nybooks.com/articles/2007/07/19/our-biotech-future/ (accessed 30 October 2017).

Fitzgerald, D., & Callard, F. (2015). Social science and neuroscience: Beyond interdisciplinarity. *Theory, Culture and Society*, 32(1), 3–32.

Foucault, M. (1976). *The history of sexuality, Volume 1: An introduction*. (Trans. R. Hurley). New York: Vintage Books.

Frost, S. (2016). *Biocultural creatures: Toward a new theory of the human*. Durham, NC: Duke University Press.

Galloway, A., & Thacker, E. (2007). *The exploit: A theory of networks*. Minneapolis: University of Minnesota Press.

Gane, N. (2005). The radical post-humanism of Friedrich Kittler: The primacy of technology. *Theory, Culture and Society*, 22(3), 25–41.

Gillborn, D. (2016). Softly, softly: Genetics, intelligence and the hidden racism of the new geneism. *Journal of Education Policy*, 31(4), 365–388.

Goodman, A. (2013). Bringing culture into human biology and biology back into anthropology. *American Anthropologist*, 115(3), 359–373.

Gravlee, C. (2009). How race becomes biology: Embodiment of social inequality. *American Journal of Physical Anthropology*, 139(1), 47–57.

Hansen, M. (2015). *Feed-forward: On the future of twenty-first-century media*. Chicago, IL: University of Chicago Press.

Harman, G. (2016). Agential and speculative realism: Remarks on Barad's ontology. *Rhizomes: Cultural Studies in Emerging Knowledge*, 30. https://doi.org/10.20415/rhiz/030.e10.

Hayles, N. K. (2017). *Unthought: The power of the cognitive nonconscious*. Chicago, IL: University of Chicago Press.

Hernandez, J., Riobo, I., Rozga, A., Abowd, G. D., & Picard, R. W. (2014). Using electrodermal activity to recognize ease of engagement in children during social interactions. In *Proceedings of the 2014 ACM International Joint Conference on Pervasive and Ubiquitous Computing* (pp. 307–317).

Ingold, T., & Palsson, G. (Eds.). (2013). *Biosocial becomings: Integrating social and biological anthropology*. Cambridge: Cambridge University Press.
Johnston, A. (2013). *Prolegomena to any future materialism: The outcome of contemporary French philosophy*. Evanston, IL: Northwestern University Press.
Kim, J., Snodgrass, M., Pietrowicz, M. Karahalios, K., Halle, J. (2013). BEDA: Visual analytics for behavioural and physiological data. Available at social.cs.uiuc.edu/papers/pdfs/BEDA_camera_ready_version_VAHC_kim.pdf (accessed November 2017),
Kirby, V. (2011). *Quantum anthropology*. Durham, NC: Duke University Press.
Kohn, E. (2013). *How forests think: Towards an anthropology beyond the human*. Berkeley: University of California Press.
Landeker, H. (2016). The social as signal. *The Sociological Review Monographs*, 64(1), 79–99.
Lazzarato, M. (2002). From biopower to biopolitics. *Pli: The Warwick Journal of Philosophy*, 13, 112–125.
Lupton, D. (2014). Self-tracking cultures: Towards a sociology of personal informatics. In *Proceedings of the 26th Australian Computer-Human Interaction Conference on Designing Futures: The Future of Design* (pp. 77–86). December 2–5. University of Technology, Sydney, Australia. New York: ACM.
Manovich, L. (2002). *The language of new media*. Cambridge: MIT Press.
Meloni, M. (2016). *Political biology: Science and social value in human heredity from eugenics to epigenetics*. Basingstoke: Palgrave Macmillan.
Meloni, M., Williams, S., & Martin, P. (2016). The biosocial: Sociological themes and issues. *The Sociological Review Monographs*, 64(1), 7–25.
Nafus, D. (2016). *Quantified: Biosensing technologies in everyday life*. Cambridge: MIT Press.
Noe, A. (2004). *Action in perception*. Cambridge: MIT Press.
Nold, C. (2009). *Emotional cartography: Technologies of the self.* Creative Commons: http://www.emotionalcartography.net/
Papoulias, C., & Callard, F. (2010). Biology's gift: Interrogating the turn to affect. *Body & Society*, 16(1), 29–56.
Piccolini, M., & Bresadola, M. (2013). *Shocking frogs: Galvani, Volta and the electric origins of neuroscience*. Oxford: Oxford University Press.
Platoni, K. (2015). *We have the technology: How biohackers, foodies, physicians & scientists are transforming human perception, one sense at a time*. New York: Basic Books.
Protevi, J. (2013). *Life, war, earth: Deleuze and the sciences*. Minneapolis: University of Minnesota Press.
Rose, N. (2013). The human sciences in a biological age. *Theory, Culture & Society*, 30(1), 3–34.
Ruppert, E., Law, J., & Savage, M. (2013). Reassembling social science methods: The challenge of digital devices. *Theory, Culture and Society*, 30(4), 22–46.
Sano, A., & Picard, R. W. (2013). Stress recognition using wearable sensors and mobile phones. In *Affective Computing and Intelligent Interaction (ACII), 2013. Humaine Association Conference* (pp. 671–676). Geneva: IEEE.
Satel, S., & Lilienfeld, S. (2013). *Brainwashed: The seductive appeal of mindless neuroscience*. New York: Basic Books.
Stevens, H. (2013). *Life out of sequence: A data-driven history of bioinformatics*. Chicago, IL: University of Chicago Press.
Van Ijzendoorn, M. H., Bakermans-Kranenburg, M. J., & Ebstein, R. P. (2011). Methylation matters in child development: Toward developmental behavioral epigenetics. *Child Development Perspectives*, 5(4), 305–310.
Wendt, A. (2015). *Quantum mind and social science: Unifying physical and social ontology*. London: Cambridge University Press.
Wheeler, M. (2005). *Reconstructing the cognitive mind: The next step*. Cambridge: MIT Press.

Williamson, B. (2016). Digital education governance: Data visualization, predictive analytics, and 'real-time' policy instruments. *Journal of Education Policy*, 31(2), 123–141.
Williamson, B., Kotouza, D., Pickersgill, M., & Pykett, J. (2024). Infrastructuring educational genomics: Associations, architectures, and apparatuses. *Postdigital Science and Education*, 6(4), 1143–1172.
Wilson, E. (2015). *Gut feminism*. Durham, NC: Duke University Press.
Youdell, D. (2016). Bioscience and the sociology of education: The case for biosocial education. *British Journal of Sociology of Education*. https://doi.org/10.1080/01425692.2016.1272406.
Youdell, D., & Lindley, M. R. (2019). *Biosocial education: The social and biological entanglements of learning*. New York: Routledge.

7

ECO-SENSORY ETHNOGRAPHY AND THE SURREALIST IMPULSE

Elizabeth de Freitas, Maggie Maclure and David Rousell

Introduction

Sensory experiences are newly tapped in a vastly expanded digital economy, which is reshaping our relations to place and culture, and distributing data across widespread circuits of connectivity (Ruppert et al., 2013). Sensing is now distributed across an increasingly complex array of digital devices, bodies, architectures, and built environments (Gabrys, 2016; Tironi, 2017). Digital sensors are "emplaced" within buildings, embedded in smartphones, worn on bodies, mounted on rooftops, orbiting in satellites, submerged in soil, flying through air, and connected to plant and animal life. This redistribution of sensory relations goes well beyond the quantified self (Lupton, 2016) and is occurring at a planetary scale, emblematic of what Rosi Braidotti (2013, 2019) has termed the "posthuman condition," a convergence of technological and climatological disruptions that radically destabilises our understanding of sensory capacity. Under these conditions, the digital has come to be part of a "general ecology" that commingles the "natural" and the "artificial" in new uncharted terrain (Hörl, 2018). In terms of ethnographic research, this opens onto what Clough (2009) envisages: "an infra-empiricism that allows for a rethinking of bodies, matter and life through new encounters with visceral perception and pre-conscious affect" (Clough, 2009, p. 44).

During the last decade, our work in the Manifold Lab at Manchester Metropolitan University explored the creative potentials of a range of eco-sensor technologies, including environmental sensors that detect invisible gases and particulate matter; force sensors that detect gravity and push/pull encounters; mobile 360 camera sensors that can scan three-dimensional objects and environments; wearable movement sensors that detect a body's speed, acceleration, and orientation; and biological sensors that sense unconscious shifts in bodily temperature, heart rate, blood volume pulse, and electro-dermal skin activity (EDA). Recent developments in mobile technologies, affective computing, neuro-marketing, movement, and geo-location sensors have made wearable digital devices increasingly accessible to corporations, governments, and the general public. The rapid movement of bio-eco-sensors and "hand-held" technology, from the laboratory

DOI: 10.4324/9781003538769-7

into the public domain, can be understood as a new stage in the relentless commodification of life through the algorithmic capture of sense data. By harnessing the otherwise invisible and asignifying play of affective intensity on the surface of the skin, for instance, the capture of biodata adds new subconscious dimensions to ubiquitous behavioural, geolocative, and predictive data profiling of individuals under corporate and governmental regimes. By tracking eye movements and micro-gestures, for instance, cameras capable of recording pre-conscious affect can fracture the "moment" into molecular signals, while software sorts inhuman amounts of images for patterns in behaviour. In short, there is now the dangerous possibility – and current reality – that governments and corporations are "mining" the intensive, unconscious, and affective dimensions of the body (de Freitas & Rousell, 2021).

Our engagement with eco-sensor technologies is driven by an interest in how digital devices might be repurposed to resist the clinical and pathologising discourses that have historically dominated the use of these technologies. We are interested in working with sense data as that which inheres in the relational environment (worldly sensibility) rather than the person (stimulus-response). This involves a complete retheorisation of sensors and data, which contests the stimulus-response psychological model of perception studies (see Chapter 6). The latter model, for instance, treats epidermal sensors (EDA) and the electrical conductivity of the skin as associated with intensity of affective response and the arousal of the sympathetic nervous system (Piccolino & Bresadola, 2013; Platoni, 2015). Typical scientific studies of skin conductance maintain the individual organism as the unit of analysis, attributing causality to an external stimulus triggering an internal response that materialises on the surface of the skin under controlled conditions (Hernandez et al., 2014). Our interest, on the other hand, is in the gathering of transindividual sensitivities whereby the "intensive atmosphere" carries within it the qualitative and quantitative ambivalences of worldly sensibility.

Might eco-sensory technologies, as conceptualised and deployed through a *surrealist* ethnography of intensities, offer new possibilities for reactivating atmospheric blocks of sensation, while resisting the imperialist eye of conventional ethnography? Deleuze and Guattari (1987, 1994) argue that the human senses should be reconsidered as part of an *asignifying* alliance, an eco-sensory ground on which the pyramid of representation is subsequently erected, its energy solidified into foundational "bricks" of culture and signification. The task then is to try to follow the intensities and forces that still move, or might move, within the hierarchical structures of conventional representation. Our interest in the imperceptible forces that lie beyond or beneath human consciousness brings our work into contact with ethnography's recurring engagement with surrealism, as we describe below. Our focus in this chapter is therefore on surrealist ethnographic experimentation with sensation. We engage with digital devices as ecological agents, tracking intensive flows, layering affect, invaginating power, ramifying insight, and entangling concepts through playful inversion, pursuing a kind of surreal participation that avoids de-scription and conventional causal exegesis. Our approach builds on Deleuze and Guattari's important elaboration of "sensation," and our own work which has contested the theorisation of sensation in perception studies (de Freitas et al., 2019).

We discuss here a project situated at the Whitworth Art Gallery in Manchester, UK, conducted in collaboration with the Young Contemporaries, a youth gallery programme.

The gallery was approached as a public space and sensory milieu, structured, according to convention, to elicit perception and judgements of cultural value. We treated the gallery as a space for experimenting with the relational capacities of art, pedagogy, and collective forms of sensing, mobilising sensor technologies as part of a surrealist experiment (de Freitas et al., 2022). This approach was partially inspired by the work of Media theorist Mark Hansen (2015) who uses the concept of "worldly sensibility" to describe a new "radically environmental" distribution of sensation within computational material-cultures. Following Whitehead, Hansen defines worldly sensibility as a vast causal infrastructure of sensory activity that operates above, around, through, and below the thresholds of human sense perception and consciousness (Hansen, 2015, p. 2). He argues that contemporary sensor technologies operate at speeds and scales that fall outside the bandwidth of human perception and thereby "impact the environment – including our bodily environment – before impacting … our higher-order sensory and perceptual faculties" (Hansen, 2015, p. 38). The micro-temporal speeds of digital algorithms, signal processers, camera mechanisms, and actuators operate so quickly that they literally feed sensory data "forward," into the future, before we have a chance to catch up. Akin in some ways to early *surrealist* experiments with the non-conscious and autonomous tendencies of events, Hansen (2015) focuses on how digital sensors fold time in new ways, opening onto more-than-human temporalities. As sensory technologies stretch and package time below thresholds of human consciousness, we see the emergence of new forms of participation, labour, exploit, control, and surplus value arising within these "augmented" posthuman milieus (Massumi, 2018). This algorithmic capture of sensation to serve corporate, managerial, and governmental agendas makes critical and creative engagements with sensors increasingly urgent, a task that our project took up through a "remixing" of sensory technologies and ethnographic surrealism.

Ethnographic surrealism

Through its periodic encounters with surrealism, anthropology has found techniques for engaging with sensation, affect and more-than-human relations that exceed interpretation and rationality. Schwanhäußer and Wellgraf (2015) identify ethnographic surrealism as a "minor" line or "rogue tradition" in anthropology. They sketch the cyclical appearance of this surrealist impulse, from its inception in France in the 1930s, when the boundaries among surrealism, anthropology, and art were not yet consolidated (Clifford, 1981).

Clifford (1981) summarised the political project of early surrealism thus: "The sort of normality or common sense that can amass empires in fits of absent-mindedness or wander routinely into world wars is seen as a contested reality to be subverted, parodied and transgressed" (p. 117). It is important to note, however, the complex relation that surrealism has always held to colonialism (Antle, 2015) and to issues of race and gender (Keeling, 2019). On the one hand, surrealist experimental projects responded directly to the threat of fascism and the subjugation of non-Western peoples, and surrealism continues to be part of decolonising and resistance aesthetics (Rousell et al., 2020). There is a new appreciation for the vast international proliferation of surrealist projects in Africa, Asia, South America, the Caribbean and Eastern Europe in the early 20th century, as a form of political resistance to all kinds of totalitarian states. Nevertheless, many critics have noted how surrealist opposition to colonialism often also replayed colonialist tropes, betrayed ambivalent attitudes to women, and adopted a problematic and exclusionary relationship to so-called primitivism (Rosemont & Kelley, 2009).

Figures such as Michel Leiris operated across the boundaries of these disciplines: colonial explorer in the Dakar-Djibouti mission of 1931–1933, contributor to the influential Surrealist journal Documents, author of the literary-ethnographic hybrid Phantom Africa, and member of the College de Sociologie, led by Bataille and Callois, which aimed to systematically document communal practices that reveal "the sacred in everyday life." Schwanhäußer and Wellgraf trace the subsequent appearances of ethnographic surrealism through its "flirtatious liaison" with folklore, art, and social documentary in movements such as Mass Observation in the UK in the 1950s, the counterculture of the 1960s, its reappearance as avatar of postmodern sensibility embraced by James Clifford in the 1980s, the postcolonial surrealism of Michael Taussig, and recent explorations of the uncanny workings of affect in the work of Kathleen Stewart and Brian Massumi.

Schneider and Pasqualino (2014) also recount the use of avant-garde experimental film tactics within the history of anthropology, as do Estalello & Criado (2018) and Salazar et al. (2017). Cinematic techniques have been used by anthropologists to disrupt the realist narrative, including superpositioning of images and jarring montage. Maya Deren's work (i.e. Meshes of the afternoon, 1943) foregrounds surrealist montage techniques in her study of trance-like feelings associated with ritual. The anthropologist Jean Rouch (i.e. Les Maitres fou, 1955) spoke of a "ciné-trance," which used the camera and editing to bring the spectator into what Deleuze called the time-image (see Chapter 5). These experiments were no less controversial in their reproduction of a colonial gaze, but they raised questions about how the scopic regime was structured through new media. More recently, Wanono (2014) describes how her ethnographic video work attempts to examine the role of the pixel (rather than the image) and thereby reckon with the role of software in the processing of video data.

The surrealists' fascination with inhuman affects, drugs, trance, dreams, ritual magic, and ghosts did not preclude a deep interest in science and technology. As Schwanhäußer and Wellgraf (2015) observe, their avant-garde experimentalism shared with normative science a commitment to a kind of objectivity that sought to bypass the proclivities of the authorial subject, in order to allow thought to manifest itself. Surrealism was experimental and speculative – aspiring to release the energies of the non-rational through experiments with automatic writing and photography, and innovative uses of film. Artists such as Ernst, Deren, and Duchamp, for instance, deployed the conventions of the magnetic or seismic trace in their art in order to introject duration into the static scene of representations. Lomas and Stubbs (2012) describe automatism as practised by the surrealists as pre-eminently speculative: "a research method, a set of investigative procedures that organise and govern practice but do not determine outcomes." They note how the surrealists appropriated early recording technologies, with the aim of transforming themselves into "modest recording devices" capable of providing a graphic trace of "forces and phenomena that do not themselves belong to a visual order of things."

At the start of our project, the Whitworth Art Gallery was featuring Thick Time, an exhibition of five complex installations by South-African contemporary artist William Kentridge. Kentridge's bold architectural modification of the gallery radically transformed the museum space from an environment that traditionally conveys deference to dominant/hegemonic cultural values. The exhibition created a distinctive multi-sensory ecosystem of multi-channel sound, film, animation, sculptural, choreographic, and architectural components in a densely affective atmosphere. The artworks relayed disparate sounds, chants, atonal music, and polytemporal rhythms through non-linear imagery and obliquely positioned megaphones, altering the visual-sonic fabric of the space continuously. Rooms were also occupied by strangely altered antique media devices, large kinetic breathing and banging machines, which maintained a consistent pulse of rhythmic activity dissonant with surrounding audio and video screenings.

Our aim was to explore young people's sensory engagement with the gallery space during six workshops and an installation. The project attempted to mobilise a postcolonial surrealist impulse that we sensed in Kentridge's installation work, whose playful dynamic installations lured the eye, only to blast apart any unity of sense in the bombastic reverberating activity and incessant movement. His installations wove together re-imagined histories of colonialism with the philosophy of art, cinema, science, and technology. Kentridge (2014, min 12) has explained this approach simply as the way of going through the world, of going through the world making sense. There is no other way. "We don't have complete information, and we cannot take it in." Kentridge's work refuses the clarity of a single or coherent narrative of both coloniality and cultural revolution and produces monstrous and awkward machines, forging a polytemporal mix of imagery, sounds, and materials.

Remixing thick time

Our workshops brought about 20 young people together for three hours at the gallery on each of six occasions. We were focused on the ways in which new media might be mobilised to alter the sense of time and participation, and how this altered form of engagement was linked to the themes and methods of Kentridge's installation of *Thick Time*. We sought to experiment with various technologies, with the aim of sensing the affective atmospheres of the exhibition, plugging bodies into the liveliness, intensity, and multi-temporality of the artworks. More generally, we were interested in the sensory dimension of gallery experiences, noting that the senses become attuned to shifting compositions of light, sound, space, colour, movement, and form. We were also interested in how gallery time can feel surreally suspended, altered, slowed, quickened, or transfixed as bodies become affected by the particular atmosphere sustained within the built environment.

Our experimental techniques in the workshops included: (1) *The haptic eye*, which involved the use of hand-held and chest-mounted video-cameras as drawing tools for tracing the movement of Kentridge's art-media machines. These videos were subsequently cut into still images at half-second intervals, printed on paper, then traced onto wax paper, photographed, and reassembled as a form of stop-motion video. (2) *The sensing body*, which involved wearing biosensors that captured changes in electrodermal activity, relayed via Bluetooth to a laptop. Participants gestured while simultaneously watching the graphical rendering of the five biometrical responses in different colour codes on the computer screen. Later, these devices were installed as part of a "concept activation" game, which further disrupted the biometric signals. (3) *Sounding time*, in which stereophonic headphones were plugged into a binaural audio-recorder and worn by participants as they circulated through the gallery space, multiplying and magnifying the tonalities and sonic atmospheres. (4) *Fugitive whispers*, which involved selecting particular places in the gallery, huddling in a small group, and then reading aloud, in a whisper, poetic texts about sensation and time (Figure 7.1).

Each of our workshops began by spreading "concept cards" out on the table and inviting young people to pick one, including "affect," "sensation," "sense," "atmosphere," "time," "attunement," and "intensity." Discussions involved sense-making and sense-breaking, to the extent that a concept such as "attunement" might travel from initial associations with "listening," "connection," and "togetherness" into much darker associations with "surveillance," "predation," and "death." After a few visits, the skin conductance devices (EDA) were introduced as playful gadgets, and different participants tried them on, and discussed how they operated. The EDA data generated during the discussion was then projected onto a screen in front of the group. This enabled everyone to witness the organic-electrical graphism, as it modulated over time, the group data drifting in and out of synchrony; ostensibly folding biosensory data back into the ethnographic site in an iterative feedback loop (see de Freitas & Rousell, 2021, for more on synchrony).

Drawing on the workshop experiments, we developed a set of installations, as part of a culminating public event at the gallery, titled Remixing Thick Time. For instance, the recorded "fugitive whispers" made during the workshops were installed into small containers evocative of travel and departure, themselves installed within a dark inflatable dome of silence within the gallery. Exhibition visitors crawled into the dome with flashlights, seeking out these containers in the dark and straining to hear these recorded whispers. We wanted to emphasise how whispers resist the urge to broadcast and refuse to fill the space, enticing bodies to come closer and stretch their senses beyond what might feel comfortable. Whispering in the gallery became a politics of counter-speech – quiet, intimate, proximal, close, secretive, opaque, not to be heard by all, without offering the comfort of contextual information, clarity of meaning or an identifiable speaker. We also created a "Concept activation game," which included a set of cards with the key concepts

FIGURE 7.1 General view of remixing thick time exhibition.

on one side, and a set of surrealist commands on the flip side, offered as "activations" of the concept. The game was meant to trouble ideas about stimulus-response in the sensory environment, destabilising the relationship among language, body, and milieu. To simulate the feeling of a game, we designed an embossed tablecloth, a set of game-like instructions, eggshell timers for tracking time on task, and arranged the set of five EDA sensors on the table, with a projector streaming live biodata from the sensors (when worn) onto the gallery wall (Figure 7.2).

People wandered into the great hall as part of their exploration of the gallery, and many people sat down at the table and began to play the game in a spontaneous, ad hoc fashion, while one member of the team – dressed in a lab coat – facilitated with the EDA sensors and helped clarify the instructions. Others stood by and watched the game play, commenting on the undulating waves of data projected onto the wall. As people started to activate concepts, enormous amounts of laughter were heard throughout the room. Each concept came with commands, which provoked the players to attend to the surreal relationship between sense and sensation, concept and meaning, language (as command) and action, underscoring the ways in which causal agency is imagined or speculative. We were able to problematise language as timed demand, "order words" and "pass-words" (Deleuze & Guattari, 1987), while situating the group as parlour-table spectators of their own playful layering of embodied rhythms.

Here "attunement" becomes a kind of unwilled affirmation, a paradoxical move to form the affirmative word "yes" when your tongue is unfree (Figure 7.3), while the rules invite breaking the rules (Figure 7.4) and "atmosphere" has visitors fogging their mobile phone screens with their own breath, before taking fuzzy pictures of the hand of someone nearby (Figure 7.5).

FIGURE 7.2 Playing the concept activation game (designed by de Freitas, Trafi-Prats & Rousell).

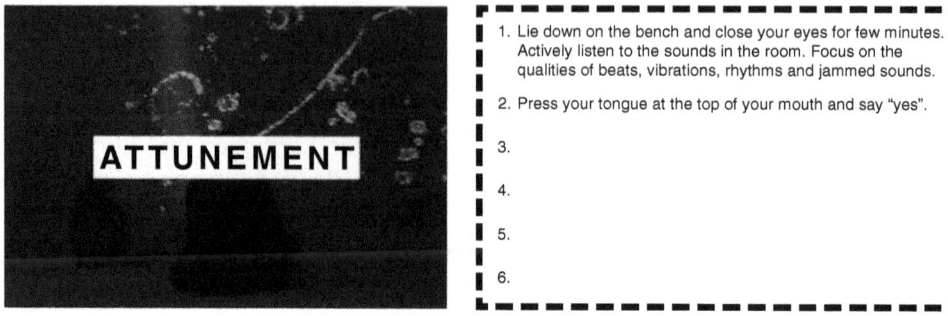

FIGURE 7.3 Concept card.

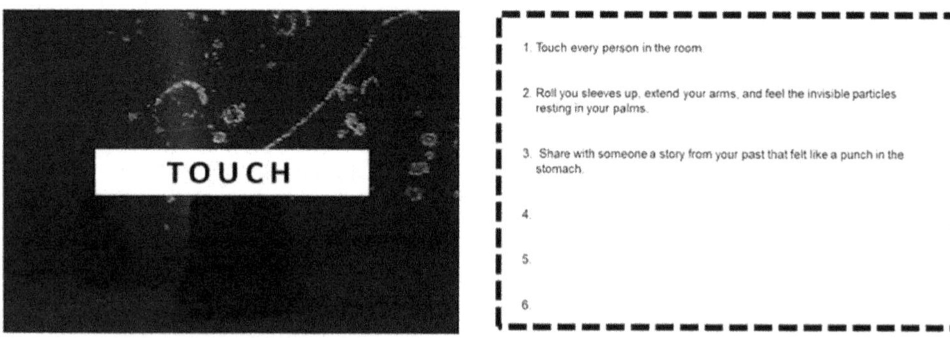

FIGURE 7.4 Concept card: touch.

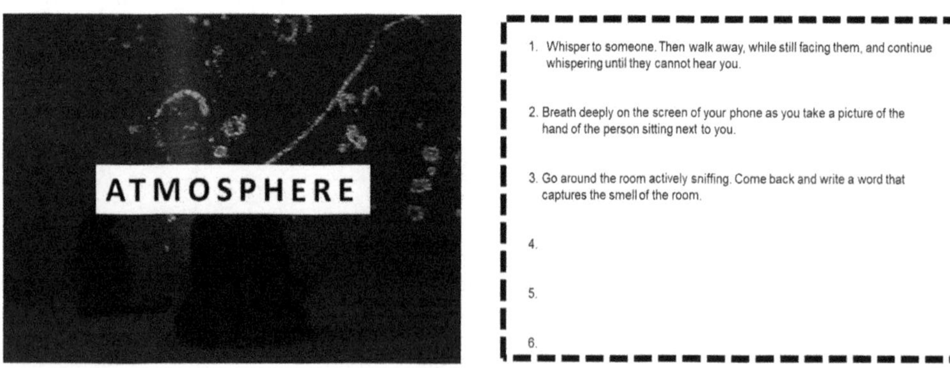

FIGURE 7.5 Concept card: atmosphere.

As one plays and performs the concept activation, the graphical output "draws" the quivering charged periphery of the conjoined sympathetic nervous system for all players. In other words, the "measure of arousal" becomes a collective expression of the affective ecology. The EDA sensory graphic fuses the five distinct signals, while smudging the gap between collective consciousness and collective sensation. Rather than framing this data as representing intensities that belonged to the individual human, the surrealist game scatters the signal into play mode. We intended for our approach to resonate with Kentridge's darkly humorous exploration of the ways in which machinic media are mobilised in the making and unmaking of memory and history. The Conceptual Activations game used biosensors as atmospheric vital media to plug into intensive flows, exploring the surrealist underside of sensory data. Our game steals back the biodata and folds it into a graphical rendering that both affirms the body as an electric-organic creature and yet undermines any claims to represent the individual as the singular agent of their sense and sensation.

Perhaps this playful event shows how ethnographic participant observation could become something akin to Stengers' "immanent discrimination" (2018) or Cull's (2011) "ontological participation," in which the opposition between participation and observation is dissolved or displaced, in favour of immersive, multi-sensory practices of attention to "our ontological participation in immanence, change and movement" (Cull, 2011, p. 91). Reading Deleuze's philosophy of immanence in relation to the "Activities" of performance artist Allan Kaprow, Cull suggests that ontological participation demands a kind of "attention training" in order to sense the imperceptible movements of difference that compose events and subjects. Kaprow's Activities ask participants to enact seemingly banal or pointless actions; for instance:

> wetting a finger with saliva
> holding it up till dry
> calling out: now!
> Walking on
> Kaprow, quoted in Cull 2011.
> (p. 80)

Our Activations share with Kaprow's Activities the aim of bypassing the assurance of the discrete subject and its continuous calibration of meaning, purpose, agency, and self-certainty, in order to attend to the quiver of the inhuman. The notion of attention training amplifies Ingold's (2014) rethinking of participant observation as an "ontological commitment" and "an education for attention" to the indescribable and the inexplicable in human affairs (p. 388).

The missing half second

While strangers sat down to play the concept activation game, we projected a film titled The Missing Half Second onto another wall of the gallery and arranged a series of beach chairs for viewing. The film was composed in our workshops with the Young Contemporaries, using short videos made with the participants, at the Kentridge show. These short videos were later split into 120 half-second still images. These images were printed and brought back to a subsequent workshop, where the images were redrawn as black and white drawings on semi-transparent paper. These 120 redrawn images were then photographed and recomposed in layers and reanimated as a moving image, smudging the temporality of the event, to foreground the rich multiplicity and ghosting of the present moment. The film explores the idea of a missing temporal interval between sensation and conscious perception (Hansen, 2006; Massumi, 2002), by continuously interleaving and superimposing the disparate images.

The film is approximately nine minutes in duration and can be viewed online at https://vimeo.com/326813533. The original video documents a participant as she herself videotaped a strange archaic set of machinic gears placed on top of a large wooden tripod mounted by Kentridge in the art gallery, moving her own camera in a circular fashion to animate the potential movements of this antique machine. As we watch the film – 120 drawn stills composed into a continuous sequence – we find that the gears of Kentridge's machine, the lens of the camera, and the face, arms, and hands of the young woman become indistinct (see Figure 7.6, top). They begin to circulate as an atmospheric body – the young woman's body disappears and reappears, slowly transformed into fragments of gears, fingers, wheels, etc. (Figure 7.6, bottom).

The process of making this film – sampling and remixing media – rendered a sense of being caught up in a maker's loop, which kept us switching back and forth between the still image and the moving image. Each of the original 120 drawn stills also lay along the gallery wall, on benches. The video was a moving image which relayed and proliferated the many layers of artistic engagement, re-installing the original Kentridge installation, as though we had contaminated the gallery with the layerings of these cut-up and re-assembled images. The Missing Half-Second film was further relayed through a reprogrammable sound system, which invited audiences to remix samples of sounds recorded by young people in the gallery, including the slamming sound of old machinery and the singing sounds of South African colonial-era music from the Kentridge show.

The use of the camera to follow the loudly slamming movements of industrial technology, and our collaborative subsequent redrawing of stills selected by the automated software, underscores how one can trouble ethnographic participant-observation, not by disabling vision per se, in a literal self-blinding; rather, it could involve an altered concept of digital vision, within ethnographic practices of sensing that are polyvocal, and synaesthetic, a vision in touch with all of itself at once (Grosz, 2017; Ruyer, 2016). Several authors have arrived at the notion of haptic vision (Deleuze, 2003; Marks, 2002; Massumi, 2011; Puig de la Bellacasa, 2009) – a form of immanent sensing that has renounced the eye's transcendent position in the attempt to apprehend the abstract forces and intensities that compose events, destabilising standard ontological assumptions about the nature of perception (Cull, 2011). Haptic vision works synaesthetically across the boundaries that arbitrarily partition sensing into discrete senses and elevate some to more prominence than others. It also operates through registering rather than representation. In Hayward's (2010) term, haptic vision might be a matter of seeing with *fingery eyes*.

We mobilise here the notion of "seeing without reading" sketched by Deleuze and Guattari in Anti-Oedipus as a practice of/for immanent or ontological participation (Cull, 2011). The eye that attempts to see without reading tries to avoid subjecting participants – human and non-human, virtual and actual – to observation's ambition to comprehend, and thereby consume, its objects. Seeing without reading is an ambulant practice (Deleuze & Guattari, 1987) – a matter of attentively following and gathering the lines and contours of abstract yet intensive flows. Paradoxically, seeing without reading always demands a jump from the known to the unknown (Deleuze & Guattari, 1983; James, 1996; Stengers, 2008). It involves acts of errant attunement, surreal disjunction, relayed attention, intensification and playful modulation. Seeing without reading implies that participation is fundamentally a matter of undergoing disjunctive movements of difference, and that participation "comes first" before perception.

FIGURE 7.6 The *Missing Half Second.* Animated video. Rousell, de Freitas, Trafi-Prats, Still images #40–45 layered (top) and still images # 66–71 layered (bottom).

We encounter this film – The Missing Half Second – as it disperses the eye across a collection of drawn fragments, reassembled and transformed into the errant graphical trace of the haptic eye. Leaning on anthropological studies of the relation of voice to the marking of the body (which they term a primitive "graphism") in ritual and healing practices, Deleuze and Guattari assert that the eye was once a confederate in a "magic triangle" of voice, body, and eye: its function was not to perceive as representation but to register the relational milieu. Similarly, voice was not signifying, but rather oracular and declamatory;

bodies similarly procured inscriptions of a localised graphism that was not yet writing but might still consist of "a dance on the earth, a drawing on a wall, a mark on a body" (1983, p. 188). The contribution of the eye was to grasp the "terrible significance" of the asignifying relations among sensations. In order to do this, Deleuze and Guattari write, "the eye jumps." This jump establishes connections that are promiscuous, mobile, radial, and above all, asignifying. Deleuze and Guattari (1983) write:

> the chain of territorial signs is continually jumping from one element to another; radiating in all directions; emitting detachments wherever there are flows to be selected; including disjunctions; consuming remains; extracting surplus values; connecting words, bodies, and sufferings, and formulas, things and affects; connoting voices, graphic traces, and eyes, always in a poly-vocal usage.
>
> *(p. 204)*

Our software decomposition of the original gallery video into half-second intervals, the ink-jet printing, and then the detailed collective labour of redrawing each image on wax paper and then photographing it, broke up the continuity of the moving image, only to then force a new unstable alliance among all these perspectives, by re-pixelating them, and using software again to achieve a soft merging and animated effect. This technique subordinated the representational quality of the ethnographic data, scrambling the camera eye into a more atmospheric registering, "a way of jumping that cannot be contained within an order of meaning, still less within a signifier" (Deleuze & Guattari, 1983, p. 204). Our surrealist technique celebrated "vision" as an ecological or atmospheric feature of an event, a kind of indigenising that forges unstable alliances or "unnatural nuptials" (Deleuze & Guattari, 1987). This experiment underscores how sensation is not just about meaning, but efficacy – that is, effecting transformation. Our technique eschews the conventional coupling of sensation-proposition-explanation and tries instead to follow the abstract lines and intensities that connect events on the virtual plane, in a process of jumping that nevertheless keeps its feet grounded in what actually occurs, as the errant expression of those intensities.

Concluding comments

We take it as axiomatic that ethnographic practices are inherently problematic, linked as they are to "colonial specularity" (Bhabha, 1994, p. 122) and its collusion in the "humanising" projects of anthropology (Atkinson, 1990; Malinowski, 1922/2005), even in the recent "ontological turns" towards animism (Povinelli, 2016; Strathern, 2020) and pluriversality as ethnographic concepts (Blaser & de la Cadena, 2018; Escobar, 2020). The yoking of sensation to participation through multisensory ethnography can be understood as an attempt to democratise the asymmetrical relations imposed by the scopic regime of colonialism (Pink, 2015). But we would argue that, despite these precautions, sensory ethnographic practice has tended to preserve the "panoptic immunity" of the liberal subject (MacLure, 2011), who exercises the prerogative to interpret the sensations of self and others by rendering them visible, transparent, and positioned without reciprocal obligation. Kirby (2011) also seeks to recast "the question of the anthropological – the human – in a more profound and destabilising way than its disciplinary frame of reference will allow." In other words, if sensory ethnography has a future as a posthumanist methodology, it may have to pass through the disciplinary thresholds of "description" and "interpretation" and become unrecognisable to itself, entering into new zones of confluence with experimental design research, non-rational modes of knowing, and speculative fabulation.

Mobilising the thought of Gilles Deleuze and Felix Guattari, this chapter pursued the ramifications of an ontological reconfiguring of sense and sensation, exploring surrealist practices of ethnography. In keeping with anthropological work associated with the ontological turn (e.g. Holbraad et al., 2014; Salazar, 2020; Salmond, 2014; Vivieros de Castro, 2004), we see this work as necessarily entailing transformation, deformation, and creative experimentation. We recruit digital technologies as confederates in this experimental work, in collaborative interventions in which humans, digital sensors, touch screens, video images, and physical artefacts are allied to bring forth flows of affect that engage and disrupt simplistic notions of participation and observation. This involved deforming or hacking or simply repurposing digital technologies, severing them from their predominant uses (such as surveillance, remediation, or the tracking of human emotions or intentions) in order to harness their potential to bring forth new ethnographic insight into gallery experiences. Drawing specifically on Deleuze and Guattari's Anti-Oedipus (1983), we reconceptualise ethnographic writing as "seeing without reading," a performative graphism that bypasses the mediations of explanation, description, and interpretation and aims to intervene directly in flows of eco-sensory affect. Language here strikes the body as surrealist command, electrical pulse, whisper, and contracted memory.

Our gallery experiments focused on the ways in which human sensations are caught up in, and transformed by, new forms of media, engendering new kinds of temporality and presencing. By engaging with the monstrous media machines of Kentridge's installations, our ethnographic "account" explores the potential of digital technology to open up "participation without observation" (Grosz, 2017; Ruyer, 2016) and "seeing without reading" (Deleuze & Guattari, 1983), where bodies participate in a mutually consequential field of sensation.

We suggest that posthuman ethnography might revisit the surrealist impulses that have periodically animated anthropology, so as to map the circulation of intensive flows and material affect without submitting to the scopic regime of representation. The microtemporal speeds of the digital and sensory excitations that we have mobilised in our museum activations could be thought of as speculation-made-concrete – attempts to "feed forward" (Hansen, 2015) by folding and stretching time into futures that encompass us but remain uncanny, because they are never solely within our human grasp. We wish to emphasise, however, the distinctive political stakes of this theoretical move for an anthropology of futures, to the extent that State and corporate interests are increasingly using both speculative fabulation (Keeling, 2019) and sensory technologies (Williamson, 2020) to prefigure "preferred" futures (Braidotti, 2017). Anthropologists and social scientists are enlisted in these efforts, and thus we see the need for a new convergence among critical ethnography, experimental art, and speculative design to reclaim sensory data from regimes of State and corporate control.

References

Antle, M. (2015). Dada and surrealism faced with colonialism. *South Central Review*, 32(1), 116–119.

Atkinson, P. (1990). *The ethnographic imagination*. London: Routledge.

Bhabha, H. K. (1994). In a spirit of calm violence. In G. Prakesh (Ed.), *After colonialism* (pp. 326–344). Princeton, NJ: Princeton University Press.

Blaser, M., & de la Cadena, M. (2018). Pluriverse: Proposals for a world of many worlds. In M. de la Cadena & M. Blaser (Eds.), *A world of many worlds* (pp. 1–22). Durham, NC: Duke University Press.

Braidotti, R. (2013). *The posthuman*. Cambridge: Polity Press.

Braidotti, R. (2017). Critical posthuman knowledges. *South Atlantic Quarterly*, 116(1), 83–96.

Braidotti, R. (2019). A theoretical framework for the critical posthumanities. *Theory, Culture & Society*, 36(6), 31–61.

Clifford, J. (1981). On ethnographic surrealism. In J. Clifford (Ed.), *The predicament of culture: Twentieth-century ethnography, literature and art*. Cambridge, MA: Harvard University Press.

Clough, P. T. (2009). The new empiricism: Affect and sociological method. *European Journal of Social Theory*, 12(1), 43–61.

Cull, L. (2011). Attention training, immanence, and ontological participation in Kaprow, Deleuze and Bergson. *Performance Research*, 16(4), 80–91.

de Freitas, E., & Rousell, D. (2021). Atmospheric intensities: Skin conductance and the collective sensing body. In J. Fritsch, B. Thomsen, & J. Kofoed (Eds.), *Affects, interfaces, events* (pp. 221–241). Lancaster, PA: Imbricate! Press.

de Freitas, E., Rousell, D., & Jager, N. (2019). Relational architectures and wearable space: Smart schools and the politics of ubiquitous sensation. *Research in Education* [special issue on "biosocial imaginaries in education"], 107(1), 10–32.

de Freitas, E., Trafi-Prats, L., Rousell, D., & Hohti, R. (2022). A poetics of opacity: Towards a new ethics of participation in gallery-based art projects with young people. In L. Trafi-Prats, R. Fendler, & A. Varela (Eds.), *Becoming lively in the city: Essays on the ontology, aesthetics and ethics of visual participatory arts-based research* (pp. 126–142). London: Routledge.

Deleuze, G. (1994). *Difference and repetition*. (Trans. P. Patton). New York: Columbia University Press.

Deleuze, G. (2003). *Francis Bacon: The logic of sensation*. Minneapolis: University of Minnesota Press.

Deleuze, G., & Guattari, F. (1983). *Anti-Oedipus: Capitalism and schizophrenia*. (Trans. R. Hurley, M. Seem, & H. R. Lane). Minneapolis: University of Minnesota Press.

Deleuze, G., & Guattari, F. (1987). *A thousand plateaus: Capitalism and schizophrenia*. (Trans. B. Massumi). Minneapolis: University of Minnesota Press.

Escobar, A. (2020). *Pluriversal politics: The real and the possible*. Durham, NC: Duke University Press.

Estalella, A., & Criado, T. S. (Eds.). (2018). *Experimental collaborations: Ethnography through fieldwork devices* (Vol. 34). New York: Berghahn Books.

Gabrys, J. (2016). *Program earth: Environmental sensing technology and the making of a computational earth*. Minneapolis: University of Minnesota Press.

Grosz, E. (2017). *The incorporeal: Ontology, ethics, and the limits of materialism*. New York: Columbia University Press.

Hansen, M. (2006). *Bodies in code: Interfaces with digital media*. New York: Routledge.

Hansen, M. B. (2015). *Feed-forward: On the future of 21st century media*. Chicago, IL: University of Chicago Press.

Hayward, E. (2010). Fingeryeyes: Impressions of cup corals. *Cultural Anthropology*, 25(4), 577–599.

Hernandez, J., Riobo, I., Rozga, A., Abowd, G. D., & Picard, R. W. (2014). Using electrodermal activity to recognize ease of engagement in children during social interactions. In *Proceedings of the 2014 ACM International Joint Conference on Pervasive and Ubiquitous Computing* (pp. 307–317). New York: ACM.

Holbraad, M., Pedersen, M., & Viveiros de Castro, E. (2014). The politics of ontology: Anthropological positions. *Cultural Anthropology Online*. Available at https://culanth.org/fieldsights/462-the-politics-of-ontology-anthropological-positions (accessed 13 January 2014).

Hörl, E. (2018). General ecology. In R. Braidotti & M. Hlavajova (Eds.), *Posthuman glossary* (pp. 172–175). London: Bloomsbury.
James, W. (1996). *Essays in radical empiricism*. Lincoln, NE: University of Nebraska Press.
Keeling, K. (2019). *Queer times, Black futures*. New York: New York University Press.
Kentridge, W. (2014). How to make sense of the word. Interview. Filmed [2014]. YouTube video, 30:24. Posted [Oct. 2014]. Available at https://www.youtube.com/watch?v=G11wOmxoJ6U.
Kirby, V. (2011). *Quantum anthropologies*. Durham, NC: Duke University Press.
Lomas, D., & Stubbs, C. B. J. (2012). *Simulating the marvellous: Psychology–surrealism–postmodernism*. Manchester, UK: Manchester University Press.
Lupton, D. (2016). The diverse domains of quantified selves: Self-tracking modes and dataveillance. *Economy and Society*, 45(1), 101–122.
MacLure, M. (2011). Qualitative inquiry: Where are the ruins? *Qualitative Inquiry*, 17(10), 997–1005.
Malinowski, B. (1922/2005). *Argonauts of the Western Pacific*. London: Routledge.
Marks, L. U. (2002). *Touch: Sensuous theory and multisensory media*. Minneapolis: University of Minnesota Press.
Massumi, B. (2002). *Parables for the virtual: Movement, affect, sensation*. Durham, NC: Duke University Press.
Massumi, B. (2011). *Semblance and event: Activist philosophy and the occurrent arts*. Cambridge: Massachusetts Institute of Technology.
Massumi, B. (2018). *99 Theses on the revaluation of value: A postcapitalist manifesto*. Minneapolis, MN: University of Minnesota Press.
Piccolino, M., & Bresadola, M. (2013). *Shocking frogs: Galvani, Volta, and the electric origins of neuroscience*. Oxford, UK: Oxford University Press.
Pink, S. (2015). *Doing sensory ethnography*. New York: Sage.
Platoni, K. (2015). *We have the technology: How biohackers, foodies, physicians & scientists are transforming human perception, one sense at a time*. New York: Basic Books.
Povinelli, E. (2016). *Geontologies: A requiem to late liberalism*. Durham, NC: Duke University Press.
Puig de la Bellacasa, M. (2009). Touching technologies, touching visions. The reclaiming of sensorial experience and the politics of speculative thinking. *Subjectivity*, 28(1), 297–315.
Rosemont, F., & Kelley, R. D. (2009). Introduction: Invisible surrealists. In F. Rosemont & R. D. Kelley (Eds.), *Black, brown, & beige: Surrealist writings from Africa and the diaspora* (pp. 1–20). Austin: University of Texas Press.
Rousell, D., Hohti, R., MacLure, M., & Chalk, H. (2020). Blots on the Anthropocene: Micropolitical interventions with young people in a university museum. *Cultural Studies – Critical Methodologies*, 21(1), 27–40.
Ruppert, E., Law, J., & Savage, M. (2013). Reassembling social science methods: The challenge of digital devices. *Theory, Culture, Society*, 30(4), 22–46.
Ruyer, R. (2016). *Neofinalism*. (Trans. A. Edlebi). Minneapolis: University of Minnesota Press.
Salazar, J. F. (2020). Speculative fabulation: Researching worlds to come in Antarctica. In J. F. Salazar, S. Pink, A. Irving, & J. Sjöberg (Eds.), *Anthropologies and futures: Researching uncertain and emerging worlds* (pp. 151–170). New York: Routledge.
Salazar, J. F., Pink, S., Irving, A., & Sjöberg, J. (Eds.). (2017). *Anthropologies and futures: Researching uncertain and emerging worlds*. New York: Routledge.
Salmond, A. J. M. (2014). Transforming translations (part 2): Addressing ontological alterity. *HAU: Journal of Ethnographic Theory*, 4(1). https://doi.org/10.14318/hau4.1.006.
Schneider, A., & Pasqualino, C. (2014). *Experimental film and anthropology*. London: Bloomsbury Academic.
Schwanhäußer, A., & Wellgraf, S. (2015). From ethnographic surrealism to surrealist ethnographies. *Reconstruction: Studies in Contemporary Culture*, 15(3), 3.
Stengers, I. (2008). A constructivist reading of 'Process and Reality'. *Theory, Culture & Society*, 25(4), 91–110.

Stengers, I. (2018). The challenge of ontological politics. In M. de la Cadena & M. Blaser (Eds.), *A world of many worlds* (pp. 83–111). Durham, NC: Duke University Press.

Strathern, M. (2020). *Relations: An anthropological account*. Durham, NC: Duke University Press.

Tironi, M. (2017). Regimes of perceptibility and cosmopolitical sensing: The earth and the ontological politics of sensor technologies. *Science as Culture*, 27(1), 131–137.

Viveiros de Castro, E. (2004). Perspectival anthropology and the method of controlled equivocation. *Tipití: Journal of the Society for the Anthropology of Lowland South America*, 2(1). Available at: https://digitalcommons.trinity.edu/tipiti/vol2/iss1/1.

Wanono, N. (2014). From the grain to the pixel, aesthetic and political choices. In A. Schneider & C. Pasqualino (Eds.), *Experimental film and anthropology* (pp. 183–198). London: Bloomsbury Press.

Williamson, B. (2020). New digital laboratories of experimental knowledge production: Artificial intelligence and education research. *London Review of Education*, 18(2), 209–220.

8

ATMOSPHERIC DATA AND SOFTWARE ARTS

New ways of investigating learning environments

Elizabeth de Freitas and Laura Trafí-Prats

Introduction

Atmospheres are ephemeral, diffuse, dissipative, "airy" and intensive. But they are also highly visceral, emplaced, tangible, affective, and material (Bille et al, 2015; Bisell, 2010). An atmosphere within a building forms a volatile space or metamorphic zone, conditioned by material architecture and the building's envelope (Zaera-Polo & Anderson, 2021). The envelope plays an important role in shaping the atmosphere, and in acting as a semi-porous membrane which links feelings of an inside to an outside. The envelope therefore is like an interface for those who enter buildings through various thresholds that are physical, social, and emotional. Buildings are charged environments, where the atmosphere carries thousands of little affects and percepts, where sensory engagement churns up and pumps up "feeling" in specific parts of a building (de Freitas et al., 2019). Humans enter the building and carry with them their memories, anxieties, and anticipations.

This chapter argues that posthuman social science can better study atmospheric data and the flow of affect by experimenting with sensor technologies and thinking with theories of digital contingency. After a decade of research on posthumanism and education, there are many new directions to pursue (Snaza & Weaver, 2015). Digital technologies have radically altered social science practice, altering and indeed displacing conventional qualitative methods championed since the 1940s, such as interview, survey, and ethnographic observation. In particular, the use of digital "passive data" captured by sensor technology has remixed the qualitative and quantitative, opening onto new methodologies like software studies and investigative aesthetics (Fuller & Weizman, 2021; Manovich, 2002). Taking cue from these two new fields, we respond to the postfoundational condition of qualitative research by closely engaging with the very same technical devices that are said to be a source of the field's dissolution (Ruppert et al., 2013). We advocate for projects that draw from the software arts, to do more than simply expose the bias and stupidity of software, which we take to be a fairly easy argument; our investigation goes deeper to engage the digital behind the curtain, and

DOI: 10.4324/9781003538769-8

to open up and experiment with the black box of software. We believe this allows for the most probing of political projects, as we reclaim and rewrite the meaning of digital data, wrestling it away from surveillance capitalists. We aim to show how software experiments engage with contingency, not only as lack or deficit in representation, but as an opening onto the trans-individual and pre-personal atmosphere of the built environment, which is evental and processual (Massumi, 2002).

We focus on learning environments which we take to be formed by physical, temporal and affective bodying habits, such as movements, departures, encounters, and anxieties, whereby a school building adapts and constrains the becoming of a collective student body. We also think with ideas on place and architecture from writers in cultural studies. Ellsworth (2004), Grosz (2001) and Parisi (2012), for instance, theorize links between learning, environment, and new media to explore the (1) pedagogy of place, (2) animacy of the built environment, and (3) digital media interventions. We also take inspiration from Architects Pallaasma (2012) and Perez-Gomez (2016) who offer expansive theories of synaesthetic architecture and lived architecture.

In this chapter, we discuss a 2020–2022 project that investigated sensory-affective experiences of school buildings, by (1) forging an interdisciplinary team from Education, Architecture, and Design who (2) worked collaboratively with young people and school staff, (3) using sensory ethnographic methods and mapping software to produce (4) new research methods for studying learning environments. The project findings are relevant to learning environment research (Woolner et al, 2007). This project generated new digital mapping methods that sought to raise awareness about the complex spatial practices and affective atmosphere within schools, whilst increasing young people's understanding of the built environment. We focus on data from one school, mapping the sensory and the atmospheric, and drawing from recent anthropological writing about atmosphere. This paper discusses how students studied the atmosphere by exploring both the building envelope, the interior passages, and various other architectural dimensions. We focus here on findings related to the use of (i) 360 image software and cameras, (ii) Polycam software and LiDAR scanner data, and (iii) Hyperspektiv software and GoPro video data, generated in collaboration with students as they ethnographically mapped the school. The students used the 360 cameras, iPad lidar camera scanners, and the GoPros to map the building envelope from the inside. The 360 images produced volumetric and navigable images, while the lidar scanner and the Hyperspektiv filters incited a different way of moving and visualizing the building, emphasizing the generative force of media to open up architecture to different kinds of corporeal and atmospheric experience (Ellsworth, 2004; Grosz, 2001).

Our research speaks to the posthuman concern with – and celebration of – contingency. This emphasis on contingency is particularly relevant to current cultural practices associated with digital life and the quantified self (Lupton, 2016; Platoni, 2015). We are not simply contesting assumptions about conventional ethnographic method, as others have persuasively done (St. Pierre, 2017), nor underlining the limits of research representation (Maclure, 2013), but advocating for a mode of inquiry that is (a) politically and materially engaged with digital transformations of life and culture, (b) attentive to the atmospheric and distributed eco-sensory nature of experience, and (c) explicitly focused on the role of contingency and speculation in experimental modes of empirical inquiry.

The surrounding envelope

As part of our research, we used 360 cameras to create volumetric images of the building interior. These surreal and disorienting images (Figure 8.1) were taken by a 360 camera of the open courtyard space of the school. The 360 image is constructed using two 180 lenses and software that stitches the two images together. The image simulates a volumetric space, and touching the screen and dragging a finger allows one to move through the space, virtually. There are often tiny stitching glitches where the software has to fudge or smudge in order to fashion a continuous surround image. Similar 360 images are used by architects online to offer a "virtual tour" of buildings. As one touches the screen to occupy and traverse these spaces, moving around the 3D image, there is a sense that the space has been captured in a strange timeless present, and that the moving point of view is disembodied, despite being positioned within the environment. Interacting with these images, students were asked to reflect on their feelings of confinement and claustrophobia. They frequently spoke about their desire to escape or to have opportunities to leave.

FIGURE 8.1 360 images of school atrium.

We used these images to disrupt the standard birds-eye view (or the drone view) of conventional architectural visions, which imagine the school building as a utopian space of rectilinear visibility and learning. Posthuman social science research has much to gain from experimenting with this new 360 image software, to expand techniques of data visualization and mapping, moving away from representational images that chart fixed locations in a grid, to a notion of topological *living maps* (Arènes et al, 2018). These images create a non-Euclidean layer, from which mapping experiments of the space can begin differently. As Aït-Touati et al. (2022) propose, a living map disrupts "point of view" in favor of "a resonant milieu, composed of thousands of superimpositions and actions of the beings that surround us, constantly and indefinitely produced by the movements and perceptions of those who create it" (Aït-Touati et al., 2022, p. 12).

These unusual images helped the students discuss the claustrophobic atmosphere of the school building and brought attention to the surrounding envelope. We share with Sumartojo and Pink (2019) an interest in studying atmosphere as "an aspect of the way something feels to people, a contingent and fluid outcome of our perpetually configured surroundings, sensory perceptions, subjectivities and imaginations" (p. 5), and their interest in the force of anticipation and "attunement" to atmosphere, as in the work of Edensor (2012) who argues that "atmospheric attunement… is frequently anticipated attunement" (p. 1114). Students' feelings about the environment were linked to their anticipatory feelings during their journey to school, which we also mapped in the workshops.

However, McCormack (2018) writes critically about the phenomenological assumptions that underpin some of Sumartojo and Pink's work (Pink et al, 2016), and others (Böhme, 2017), and notes that "invoking atmosphere as a shorthand for an empiricism that privileges, presence, immediacy, and immersion," is a way of reclaiming an authentic and prominently anthropocentric experience (p. 8). He adds, "even as it seems to privilege immersion, atmosphere is haunted by something in excess of immediacy: that which is withdrawn from apprehension, even when actively manipulated by devices and technologies that hold onto the process of immersion" (p. 8). McCormack underscores the importance of experiments in "envelopment" that "open up different modes of being and becoming more attuned to… the affective force of atmospheres as environmental milieus" (p. 9). We agree with McCormack (2017) that the relation between atmosphere and envelopment is extremely important and that modifying the conditions of the envelop can have implications for how different forms of life are sustained while others are untenable. Experiments that address the complexity of envelopment – in all its imagined shapes and contours – are needed to help people modify, puncture, and withdraw from atmospheres.

Distorting affect

Hyperspektiv video filters are a highly ranked AR app, which offers ways of altering and modulating video through various trippy distortions and feedback loops (https://www.hyperspektiv.com/). This software allowed students to modify the GoPro video that they made as they moved through the building, mapping their various trajectories through the space, so as to better capture the affect of corridors and stairways. Below are two images (Figure 8.2), each obtained with a filter on a student video that recorded movement along the fourth floor before descending the open stairs into the atrium. The bottom image used a filter that created an "ephemeral" effect that students selected for how it evoked

FIGURE 8.2 Hyperspektiv filters of navigation videos in the school building.

the dangerous and foreboding nature of the building, while the top image used a visual reverb (repeat) filter to create visual echoes and material streaks, essentially destabilizing the hard lines that would otherwise separate different spaces. The filtered videos create a temporal trace of that which is seen in passing, underscoring the bodily movement of the camera operator, which is translated into bumpy streaks of banister or wall.

The students used the filters to modify the material texture of the building and capture the feeling of moving through the building's atmosphere. This data may seem rather impersonal and alien, but it raises a more posthuman perspective on the environment. We use software to explore the more-than-human aspect of atmosphere, unlike Sumartojo and Pink (2019), who critique architect Zumthor (2006) for describing atmospheres in terms of collective clouds of materiality (smoke, light, smell) because they suspect this anonymizes the data across a trans-individual group. They are critical of posthuman approaches, which they claim diminish the causal power of humans, and they are concerned that studies of "affective" atmosphere tend to decenter human perception: according to Sumartojo and Pink (2019), such work wrongly treats atmospheres as ephemeral, as though detached from their material grounding. They have concerns that atmosphere will be reduced to "affective fields" of molecular affect transmitted from person to person, and "diminish the importance of individual experience, environmental, social and cultural factors, and the specific configurations of places and temporalities in which atmospheres might be felt and shared" (Sumartojo & Pink, 2019, p. 30).

Sensitive to these concerns, we yet diverge and follow Latour (2005) and Weibel (Latour & Weibel, 2005) in pursuing a more-than-human relational ontology and technical milieu that we hope does justice to these concerns. The humans in our study are central to shaping the atmosphere because they are powerfully positioned to do so, as part of this political posthuman ecology, and as participants in the technical process of abstraction that occurs in the project (de Freitas et al., 2022). Our ethnographic work refuses the humanist image of the conscious, reflexive subject while affirming the powerful agency of student participants. Students were empowered in workshops as they learned about software, and they were empowered by the ways in which such software allowed them to trouble conventions of visibility. Moreover, thinking about the atmosphere in posthuman terms helped them study affect as belonging fundamentally to environments (rather than individuals), which empowers people as belonging to a transindividual ecology. Notably, and we mention here to trouble the theoretical water, Sumartojo and Pink (2019) positively cite affect theorists Massumi and Seigworth, who are posthuman thinkers: "For example, Gregg & Seigworth (2010, p. 3) argue of affect that it is 'integral to a body's perpetual becoming… pulled beyond its seeming surface-boundedness by way of its relation to, indeed its composition through, the forces of encounter.'" (Sumartojo & Pink, 2019, p. 44). They turn to Massumi to help them stretch away from the phenomenological tradition:

> Here atmosphere offers the 'surplus of unacted-out potential that is collectively felt' that Massumi (2015: 57) links to emergent and uncertain affective intensities, the potential for new configurations, for a shift in thinking or perceiving that might make new futures possible.
>
> *(Sumartojo & Pink, 2019, p. 45)*

In the concluding section of this paper, we return to some of these theoretical delineations, after discussing one more experiment in the next section.

Scanning and texture

Sensor technologies often operate at scales of discernment that are imperceptible to humans, generating data that is pre-personal, pre-individual, pre-subjective, and pre-conceptual, and then remixing large amounts of qualia or sense data through instant visualizing software (Carpo, 2017). This opens onto what Clough (2009) envisages: "an infra-empiricism that allows for a rethinking of bodies, matter and life through new encounters with visceral perception and pre-conscious affect" (p. 44). Our project sought to work with such technology so that we might further explore the infra-sensory ecologies of school buildings and the social-electrical framings of embodiment and environment (de Freitas & Rousell, 2021; de Freitas et al., 2020).

LiDAR scanner sensors operate differently than standard video cameras; they emit pulsed light waves into the environment, which bounce off surrounding objects and return to the sensor. The time it takes each pulse to return is recorded and used to calculate the distance it travelled. This technical engagement with depth of space makes the device a kind of *haptic* camera (perhaps what Deleuze might have called a haptic-eye). When this haptic process of palpating the envelope or surface of a space is repeated thousands or millions of times per second, one is able to generate precise real-time 3D maps. The technology is similar to radar and sonar, using light waves instead of radio or sound waves. The scanner creates a "point cloud" of data, capturing the shape of objects and the depth of a room, which is then calibrated with the camera and pixel data to create an accurate rendering of the space. This kind of technology has been used in geospatial research since the 1980s. Today, it is applied in architecture to create highly accurate 3D models of buildings derived from large amounts of such data. The models are often used to retrofit the decision-making around potential building transformations, repairs, and installations. The technology is now widely available in wearable and portable devices like iPads and GoPros, although these handheld devices operate without the same accuracy. In this project, students were able to use LiDAR scanners installed in the 2020 iPad Pro computers.

These images (Figure 8.3) document a group of students scanning the environment, using the Polycam software on the iPads. After they turn on the app, the screen shows a bluish filter swiftly covered by a growing net of flickering points that suddenly reveal objects, surfaces, and fixtures. While looking at the iPad screen, one student asks about the blue holes flashing on and off around the edges of certain objects in the room: "How is it that these parts are missing?" They then realize that the LiDAR does not work like a video camera, where every object is projected onto a planar image (or a 360 image). The LiDAR is capturing depth, and the blue holes designate an absence of depth data. To resolve the glitch, they need to move with the iPad in multiple directions and speeds, to approach surfaces in order to sense the depth, texture, and volume of the space, rather than simply document static "a point of view." The depth data is indeed combined and

FIGURE 8.3 Students scanning the building.

calibrated with the conventional camera data, so two kinds of data are synchronized to make the final scanned image. The LiDAR's computational vision pursues this architectural space without a point of view, unlike conventional photography and cinema (Galloway, 2022). The LiDAR scanner's computational vision does not have one point of view, but rather enacts all points of view, like but different to the 360 images. These images were used to spur discussions with students about their sensory experience of the school (how they move, see, touch, smell, and feel about being in the building).

The software processes the LiDAR data and renders a 3D model, presenting the school space as both hyperrealist and flawed volume, replete with uncanny details but also holes, jagged edges, and broken borders (Figure 8.4). The students explored these images using common touch screen gestures such as tap, fling, pinch open and close, engaging the familiar spaces from new and unanticipated angles, glancing from queer perspectives (underneath the floor or outside the window) and similarly rare movements (newly proximate to a surface or corner). Partly enchanted, partly confused, they compared and discussed their resulting scans, acknowledging the software's capacity to compute the same model in parallel machines, while also recognizing errors, incompatible aspects and unknowns, what Parisi (2019) describes as *contingencies* overwhelming rational systems. The conversations foregrounded the turbulent multiplicity of matter's tendency towards variation and commotion (Spence, 2021, 2022). As Massumi (2002) notes, the intensity of affect resides in such atmospheric multiplicity, precisely because affect remains largely unknowable, inciting speculation and opening the space to mutation. The scanning experiment opened up the contingency of the building's envelope, its potential for

FIGURE 8.4 Polycam 3D scan model.

leaks, escape routes, lines of flight (real and imagined) from the controlling institution. This approach of working closely with residents resonates with architectural practices where models are not just generated as representations of a normative reality, but function as diagrammatic tools to propel experimental thinking (Poole, 2020).

The students used the software to index the limit of the interior space, to feel the constraining force of the building envelope, and push past conventions of encounter. As they scanned, their movement and activity altered, crouching, shuffling, crawling, and bending over, holding the iPad with their two hands, often swaying back and forth in front of the surfaces of walls, counters, windows, corners, as well as underneath tables and other negative spaces. They palpated the borders that separate inside from outside. These altered modes of engagement showed how the scanner is a device that attends and attunes to experiences of envelopment, confronting feelings of containment. This kind of software has proven essential in various spatial justice projects within the field of investigative aesthetics, where digital sensors and cameras are used to reveal buried stories about controversial events. Such projects recruit technology as part of the critical agenda: "the reorganization of relations between organic and inorganic sensorial matter, people and computers, can increase non-hierarchical sensorial assemblages and even seek to harvest or "rescue" traces from beneath layers of erasure" (Fuller & Weizman, 2021, p. 57). This attention to the forensic potential of data involves a "hyper-aesthetisation" or an amplification of the senses through the powers of interconnected technologies that function as "interlocutors" (p. 58), making "felt" a series of unexpected interplays between different modalities (digital, architectural, bodied). Fuller and Weizman (2021) explain that "to hyper-aestheticise is to heighten, elicit or exacerbate the capacity of bodies, technologies or states of matter to sense and increase perceptual experience; it entails an increase in sensitivity and can perhaps augment a capacity to care" (p. 58).

Conclusion: quivering contingency

The students used the digital technology to open up the building envelope, and undermine its constraints of movement and visibility, while documenting the "feeling" of corridors and passages. This research brought attention to the trans-individual and pre-personal atmosphere of the learning environment. The building envelope became something animate and adaptive that refused simplistic representation (Zaera-Polo & Anderson, 2021, p. 12). The atmospheric data exposed the building as an affective environment, showing how key areas were connected not through conventional passages, but through affect, imagination, and emotional attachments. These experiments with posthuman digital ethnography allowed the students to realize the manifold shape of interior school spaces, and to "sense" the atmosphere in new ways, revealing how the building envelope joins and modulates inside and outside environments (Zaera-Polo & Anderson, 2021). When we situated these images within other maps of the school and the community, students were able to link the interior affective atmosphere with larger processes of climate and history at various scales (see sample project materials at https://www.schoolatmospheres.net/).

Under current algorithmic conditions, the digital has come to be part of a "general ecology" that commingles the "natural" and the "artificial" in new uncharted terrain (Hörl, 2018). Braidotti and Fuller (2019) suggest that software processes inherit a relation to the outside, and "prehend a milieu" by incorporating cybernetic notions of loop and plasticity. Fazi (2018) argues for a "computational aesthetics" to be understood as "conceptual prehension […] a grasping, seizing, or holding of the pure potentiality of an ideal" (p. 12). She argues that computing machines harbor creative contingency within them; we need to better understand the glitch fragility of digital devices and the limits of computability. Fazi's focus on the inherent contingency of algorithms echoes similar claims from 1970s postmodernists, in particular Jean-François Lyotard, who argued in *The Postmodern Condition* about the inherent uncertainty of *any knowledge* or epistemic platform or paradigm (Lyotard, 1979). Fazi brings this claim up to date. The intrinsic incompleteness within any digital infrastructure is what makes computation always already a speculative leap into contingent activity. In other words, Fazi reminds us that contingency is not an after-effect of application but is rather *inherent* to all machinic activity. Building on this insight, our work responds to questions about pre-personal and pre-individual factors in social research, as well as the role of contingency in research method.

By taking a deeper dive into the technical abductive habits of digital devices, we can make visible the quivering contingency that is at the heart of both digital life and digital methods of inquiry. This is a kind of "investigative aesthetics" (Fuller & Weizman, 2021) that directs our attention to issues of spatial justice. With that in mind, this project uses digital technology as part of a speculative method that enlivens the school milieu, while also revealing the incomputable within computation as such (Parisi, 2017). Note that our project is both creative and critical – our experiments rely on digital methods to materialize the *incomputable* of atmospheres. Software is not simply processing information and spitting out representations (de Freitas, 2022). In other words, the software experiments *retroductively elicit* (or we might say "abduce") the material building's *incomputable potentiality*. Abduction is a process fully automated into our generative algorithms today, and explains why they have such a tendency towards delirium and fabrication (de Freitas, 2022). Our method here – software arts – elicits or abduces the inherent incomputability of environments. And this matters, because in retroductively eliciting the potentiality of the building, we use the software to help the students rethink and remake the future of their space and architecture. The software effects an abstract image that doesn't represent the building, but rather brings forth the latent potentialities in the building, that being the latent potentiality of the school and its atmospheres. The building then becomes animate within a newly configured field of ontological relations. The research method is here transformative of the object of study, but perhaps more importantly, the plasticity of the building is lured out of stasis by the student's engagement with the technical devices. Here, research becomes a kind of *software art* that exposes the affective atmosphere of the learning environment, while also creating new lines of flight.

References

Aït-Touati, F., Arènes, A., & Grégoire, A. (2022). *Terra Forma: A book of speculative maps* (Trans. Amanda DeMarco). MIT Press.

Arènes, A., Latour, B., & Gaillardet, J. (2018). Giving depth to the surface: An exercise in the Gaia-graphy of critical zones. *The Anthropocene Review*, 5(2), 120–135.

Bille, M., Bjerregaard, P., & Sørensen, T. (2015). Staging atmospheres: Materiality, culture and the texture of the in-between. *Emotion, Space and Society*, 15, 31–38.

Bisell, D. (2010). Passenger mobilities: Affective atmospheres and the sociality of public transport. *Environment and Planning D: Society and Space*, 28, 270–289.

Böhme, G. (2017). *Atmospheric architectures: The aesthetics of felt spaces*. London: Bloomsbury.

Braidotti, R., & Fuller, M. (2019). The posthumanities in an era of unexpected consequences editorial for the special issue on the Transversal Posthumanities. *Theory, Culture and Society*, 36(6), 3–29.

Carpo, M. (2017). *The second digital turn. Design beyond intelligence*. Cambridge, MA: MIT Press.

Cleveland, B., & Fisher, K. (2014). The evaluation of physical learning environments: A critical review of the literature. *Learning Environments Research*, 17(1), 1–28.

Clough, P. T. (2009). The new empiricism: Affect and sociological method. *European Journal of Social Theory*, 12(1), 43–61.

Daniels, H, Stables, A, Tse, H., & Cox, S (2019b). *School design matters*. Abingdon: Routledge.

Daniels, H., Tse, H. M., Stables, A., & Cox, S (2019a). Design as a social practice: The experience of new-build schools. *Cambridge Journal of Education*, 49(2), 215–233.

de Freitas, E. (2022). The role of abduction in mathematics: Creativity, contingency and constraint. In L. Magnani (Ed.), *The Springer handbook on abductive cognition*. Available at https://link.springer.com/referenceworkentry/10.1007/978-3-030-68436-5_34-1.

de Freitas, E., & Rousell, D. (2021). Atmospheric intensities: Skin conductance and the collective sensing body. In J. Fritsch, B. Thomsen, & J. Kofoed, J. (Eds.), *Affects, interfaces, events* (pp. 221–241). Imbricate! Press. Available at https://imbricate.press/wp-content/uploads/Affects-Interfaces-Events-CH10.pdf

de Freitas, E., Rousell, D., & Jäger, N. (2019). Relational architectures and wearable space: Smart schools and the politics of ubiquitous sensation. *Research in Education* [special issue on "biosocial imaginaries in education"], 107(1), 10–32.

de Freitas, E., Trafi-Prats, L., Rousell, D., & Hohti, R. (2022). A poetics of opacity: Towards a new ethics of participation in gallery-based art projects with young people. In L. Trafi-Prats, R. Fendler, & A. Varela (Eds.), *Becoming lively in the city: Essays on the ontology, aesthetics and ethics of visual participatory arts-based research* (pp. 126–142). Routledge.

Dovey, K., & Fisher, K. (2014). Designing for adaptation: The school as socio-spatial assemblage. *The Journal of Architecture*, 19(1), 43–63.

Edensor, T. (2012). Illuminated atmospheres: Anticipating and reproducing the flow of affective experience in Blackpool. *Environment and Planning D: Society and Space*, 30, 1103–1122.

Ellsworth, E. (2004). *Places of learning: Media, architecture, pedagogy*. New York: Routledge.

Fazi, B. (2018). *Contingent computation: Abstraction, experience, and indeterminacy in computational aesthetics*. Lanham, ML: Rowan & Littlefield.

Fuller, M., & Weizman, E. (2021). *Investigative aesthetics: Conflicts and commons in the politics of truth*. New York: Verso.

Galloway, A. R. (2022). *Uncomputable: Play and politica in the long digital age*. New York: Verso.

Gregg, M., & Seigworth, G. (Eds.). (2010). *The affect theory reader*. Durham, NC: Duke University Press.

Grosz, E. (2001). *Architecture from the outside: Essays on virtual and real space*. Cambridge, MA: The MIT Press.

Hörl, E. (2018). General ecology. In R. Braidotti & M. Hlavajova (Eds.), *Posthuman glossary* (pp. 172–175). London: Bloomsbury.

Latour, B. (2005). *Reassembling the social: An introduction to actor-network-theory*. Oxford: Oxford University Press.

Latour, B., & Weibel, P. (2005). *Making things public: Atmospheres of democracy*. Cambridge, MA: MIT Press.

Lupton, D. (2016). *The quantified self: A sociology of self-tracking*. Cambridge: Polity Press.

Lyotard, F. (1979). *The postmodern condition: A report on knowledge*. Manchester, UK: Manchester University Press.

MacLure, M. (2013). Researching without representation? Language and materiality in post-qualitative methodology. *International Journal of Qualitative Studies in Education*, 26(6), 658–667.

Manovich, L. (2002). *The language of new media*. Cambridge, MA: MIT Press.

Massumi, B. (2002). *Parables for the virtual: Movement, affect, sensation*. Durham, NC: Duke University Press.

Massumi, B. (2015). *Politics of affect*. Cambridge: Polity Press.

McCormack, D. P. (2017). Stratospheric envelopes for an atmospheric mode of address. *GeoHumanities*, 3(2), 414–430.

McCormack, D. P. (2018). *Atmospheric things: On the allure of elemental envelopment*. Durham, NC: Duke University Press.

OECD. (2017). OECD framework for a module on the physical learning environment – Revised edition. *Directorate for Education and Skills Education Policy Committee*. Available at https://www.oecd.org/education/OECD-FRAMEWORK-FOR-A-MODULE-ON-THE-PHYSICAL-LEARNING-ENVIRONMENT.pdf.

Pallaasma, J. (2012). *The eyes of the skin: Architecture and the senses*, 3rd edition. Hoboken, NJ: Wiley.

Parisi, L. (2012). Speculation: A method for the unattainable. In C. Lury & N. Wakeford (Eds.), *Inventive methods: The happening of the social* (pp. 232–244). London: Taylor & Francis.

Parisi, L. (2017). Computational logic and ecological rationality. In E. Hörl (Ed.), *General ecology: The new ecological paradigm* (pp. 75–100). London: Bloomsbury.

Parisi, L. (2019). Media ontology and transcendental instrumentality. *Theory, Culture, Society*, 36(6), 95–124.

Perez-Gomez, A. (2016). *Attunement: Architectural meaning after the crisis of modern science*. Cambridge, MA: MIT Press.

Pink, S., Horst, H., Postill, J., Hjorth, L., Lewis, T., & Tacchi, J. (2016). *Digital ethnography: Principles and practice*. New York: Sage Publications.

Platoni, K. (2015). *We have the technology: How biohackers, foodies, physicians & scientists are transforming human perception, one sense at a time*. New York: Basic Books.

Poole, M. (2020). Allography and the baroque agency of the objectile. In A. Beech & R. Mackay (Eds.), *Construction site for possible worlds* (pp. 171–193). Falmouth, UK: Urbanomic.

Ruppert, E., Law, J., & Savage, M. (2013). Reassembling social science methods: The challenge of digital devices. *Theory, Culture, Society*, 30(4), 22–46.

Spence, C. (2021). Senses of place: Architectural design for the multisensory mind. *Cognitive Research: Principles and Implications*, 46(5), 1–26.

Snaza, N., & Weaver, J. (2015). *Posthumanism and educational research*. New York: Routledge.

Spence, C. (2022). *Sense-hacking: How to use the power of the senses for happier and healthier thinking*. London: Viking Penguin.

St. Pierre, E. A. (2017). Haecceity: Laying out a plane for post qualitative inquiry. *Qualitative Inquiry*, 23(9), 686–698.

Sumartojo, S., & Pink, S. (2019). *Atmospheres and the experiential world: Theory and methods*. New York: Routledge.

Woolner, P., Hall, E., Higgins, S., McCaughey, C., & Wall, K. (2007). A sound foundation? What we know about the impact of environments on learning and the implications of Building Schools for the Future. *Oxford Review of Education*, 33(1), 47–70.

Zaera-Polo, A., & Anderson J. S. (2021). *The ecologies of the building envelope*. Hoboken, NJ: Blackwell Publishing.

Zumthor, P. (2006). *Atmospheres: Architectural environments, surrounding objects*. Basel: Birkhauser.

9

RE-ANIMATING THE BUILT ENVIRONMENT THROUGH MAPS AND MODELS

Laura Trafí-Prats and Elizabeth de Freitas

Introduction

Haraway (2013) has argued that speculative fabulation and fantasy are crucial components of all scientific inquiry, and Latour (2005, 2011, 2017) has similarly argued that science is a highly speculative practice, which is at its best when proliferating agencies in the "metamorphic zone" of political ecology. They both affirm the need to make such speculation explicit, so as to properly recognize the way that science works. Strangely, the power of speculative thinking in the *social sciences* is not as well recognized. In this chapter, we argue that the cosmic artisan working in the social sciences might solicit and synthesize speculative images and models of the world, which animate and pluralize the distributed agencies at work in the milieu. Such forms of inquiry make visible or tangible the ways in which an environment is alive with potentiality. With that aim, we discuss theory and methodology that help us design experiments in posthuman ethnography. Our interest is in investigating school buildings in the aftermath of the Covid pandemic, when students and staff have developed a raised awareness of school architecture and have internalized a great deal of anxiety about social proximity and shared space in high-density buildings. During the height of the pandemic, various new protocols were introduced to manage the movement of potentially viral student bodies, while sections of buildings and corridors were suddenly off-limits or given new functions. The historical moment offered an opportunity to study schools as transitional ecologies. We follow Kay et al. (2019) in foregrounding an ecological understanding of the school environment where it is possible to create experimental opportunities for participants to further develop a speculative relationship with school spaces. To that end, we organized workshops around *futuring practices* whereby sensory-affective ethnographic data was synthesized into alternative speculative maps and models of the building.

We focus here on some of the speculative "living maps" and 3D models that were generated in the study. These speculative visualizations and models made visible both the limits and possibilities of spatial life in school buildings. They rendered visible (and therefore thinkable) dynamic relations between bodies, things, materials, and previously

DOI: 10.4324/9781003538769-9

unexamined social-emotional forces in the school. These attempts at visualizing and modelling the affective-sensory dimension of the building draw attention to new modes of socio-technical life in the school after the Covid pandemic. We first situate the project in the UK context and theoretically frame the project in relation to spatial and architectural theory. Our methodology draws on various interdisciplinary scholars who study the built environment as a complex political ecology which includes non-human agents, drawing especially from the work of Bruno Latour.

Adopting a Latourian perspective helps us to position our contribution as not necessarily *acting against* school design, but as acting *within* the field of school design, creatively pursuing new possible spatial rearrangements of old school buildings. Futuring practices treat buildings as metamorphic and open to re-imagining. This emphasis on the metamorphic zone of lived experience in schools – what Latour might term (2018) "the Terrestrial" – allows us to highlight the power of the imagination in reshaping school buildings. Latour argues that all spatial visualizations combine some degree of *realism* with some degree of *imagination* and recognizes that speculation is an important facet of all descriptive work, be it mapping, modelling, or some other rendering (Latour, 2005, 2008, 2018; Latour & Weibel, 2020; also see Yaneva, 2022). Our research works collaboratively with young people and staff in schools, using ethnographic data to create speculative maps and models. These maps and models are both speculative and descriptive of the life of the building, and as such, they are intended to open onto multiple futures. After discussing the UK context, the theoretical framework, and the research design, we then present and explain some of the speculative maps and models.

Context

Our research involved a partnership with an international architecture firm called BDP, which had a long history of government contracts for building schools in the UK. They supplied us with various kinds of archived data about three school buildings in Liverpool. In this paper, we discuss a school building built in 2012, for one of the first established coeducational secondary schools in the Northwest of England in the late 1950s. This involved moving the school to a different site because the historical building was deemed obsolete. The new building integrated all school spaces in one building, incorporated more sustainable energy systems, innovative digital technology, large open spaces and classrooms, and community-facing spaces. Some of the staff currently working in the new building either worked or studied in the older building, and as such had a sense of the differences between the two designs. Soon after the new building opened, there were some alterations introduced in the open space plan to address noise problems, student flow problems, air quality problems, lighting problems, and even leaky roof problems that plagued the new building.

Initiatives of new school design in the last decade have emerged worldwide in countries like Australia, New Zealand, Denmark, England, and Iceland (Burke, 2013; Cardellino & Woolner, 2019; Daniels et al. 2019a, 2019b; Niemi et al., 2022; Salmarsh et al. 2015; Sigurðardóttir & Hjatarson, 2016; Wells et al., 2018). In the UK, these have been associated with government programmes like Building Schools for the Future (henceforth referred to as BSF) (2003–2010), which aimed at re-building and transforming all the secondary schools in England. Characteristic of these programmes has been the design of school buildings articulated through more open and flexible spaces for learning and socialization, where the use of IT is sought to promote collaboration and decentralization. While ideas of new school design have been part of school architecture for more than three decades (Cooper, 1981; Cuban, 2004; Deed et al., 2014), its current prominence in policies of education reform (OECD, 2017), along with the large capital investments at national level, and the impact on school spatial practices post-pandemic, highlight the need for research that can shed light on the lived experience within these school buildings (Cleveland & Fisher, 2014; Daniels et al., 2019a, 2019b; Dovey & Fisher, 2014).

In the UK, many of the studies concerning new school design have shown that these buildings have failed on various counts: they did not alter the traditional hierarchies and power relations (Burke & Grosvenor, 2008), nor addressed problematic surveillance and self-surveillance (Thorham & Myers, 2012); new school buildings normalized questionable discourses about vulnerable youth (Kraftl, 2012), and failed to advance new visions of education, space, and schooling (de Besten et al., 2011). Many of these studies have been critical of the ways neoliberal governance and the marketization of school procurement have curtailed architecture's social and conceptual potential to re-shape what school life and learning could be (de Besten et al., 2011). Some question the programme's actual commitment to futuring as a practice of opening up opportunity for creative learning (Kraftl, 2012; Kraftl et al., 2022). The Covid-19 pandemic brought increased attention to the limitations of current school design, as concerns about air quality and other invisible atmospheric characteristics took on new significance. Our project aimed to understand how students and staff might recast possible futures within these buildings.

Latour and the Terrestrial

Bruno Latour's (1947–2022) work extends over many fields, including sociology, anthropology, and philosophy, and more marginally, architecture and urban planning. His central contribution to contemporary philosophy has been through his studies of the production of scientific knowledge, questioning the epistemological divisions that modernity instituted between nature and culture, subject and object, science and politics, and the natural and the artificial. Borrowing from Albert North Whitehead and William James – and informed by his own anthropological studies – Latour devises a relational philosophy that attends to complex networks and dynamic ensembles composed of heterogeneous elements and forces (Latour, 1993).

Latour's (2005) philosophy incorporates a spatial theory, as he departs from Newtonian-Kantian metaphysics and their rendering of "a space of calculation [where] … every difference is supposed to have its place" (p. 216). He also critiques the even more trenchant tendency to imagine space in terms of containment and envelopment, noting that "[it] served to unify everything" (p. 215). He argues that the multiplication and scale of ecological crises since the 1970s, often called the Anthropocene, makes it impossible to think of the Earth in terms of a home built to house humans. The earth is a *multiple earth* (Latour, 2017, 2018), alive with fierce contingency. For Latour, we must attend to how space morphs, acts, and terraforms, in shapes that clearly escape containment, producing many simultaneous worlds and generating different space-times (Latour, 1997; see also Yaneva, 2022).

Latour (2018) proposes the concept of the *Terrestrial* with capital T to think of space as earthbound and heterogeneous. He clarifies that rather than a definite place to be occupied, the Terrestrial is "a new political actor" (p. 41). The Terrestrial reacts and turns against the accumulated effects of human intervention in the Earth systems, and forces us to think space as being "agitated" (p. 42), posing the question, "How do we occupy land if it is this land itself that is occupying us?" (pp. 41–42). Latour suggests that we must develop a more nuanced understanding of *occupy* one that responsibly highlights the metamorphic nature of the material world (Arènes et al., 2018) and the agentic nature of matter more generally (Latour, 2018).

In composing co-habitable worlds and *the multiple earth*, Latour (2018) argues that all cartographic and cosmographic projects must heed the creative work of artists, designers, and architects; such transdisciplinary work will then consist in reorienting aesthetically towards the Terrestrial, attuning to the complex vitality of matter while initiating a new encyclopaedic effort of *describing* what life is after the deep transformations enacted by modern science, capitalism, and globalization. Thus, the task of contemporary map makers is to create ways to visualize specific *ecologies* at work (Tresch, 2005). This involves attending to flow (Harris & Rousell, 2022). Ecology is a term that Latour understands as referring to relations of interdependence that are not limited to living organisms, but that also include dwellings and movements of things, machines, information, and affect. In other words, the term Terrestrial is not intended to separate the natural from the artificial, but rather includes the built environment and the many technologies embedded therein. Moreover, Latour recognizes the importance of incorporating techniques of *explicitation* alongside visualizations and descriptions of ecological relations (Latour, 2008, 2022; see also Aït-Touati et al., 2022). Explicitation is an awkward term meant to emphasize the need to *explicitly* expose the vision or worldview that is captured and conveyed in any particular cosmogram or map, revealing the limits of given representational methods (Tresch, 2020). Thus, mapping techniques together with explicitation provide descriptive knowledge of a given social-political ecology, while also surfacing the ways in which the distinct visual model serves specific perspectives and values (see also Drucker, 2021).

New practices of visualization in design and architecture

Our study is not the first in unravelling the connection of Latour's philosophy with architecture and architectural visualization. The field of architecture has a long relation with Latourian research, methods, and concepts (Yaneva, 2022), and Latour himself often collaborated with architects, designers, and artists. In this section, we discuss how techniques of description and speculation that align with Latour's orientation towards the Terrestrial have shaped new practices of visualization in architecture and design.

Architects informed by Latourian philosophy, and whose work aims to respond to new socio-material configurations in the Anthropocene, have proposed methods of visualization that map design *in process*, avoiding a vision of architecture as ever settled in place (Yaneva, 2012, 2022). They follow the actors and the actions involved in designing a building. They map, trace, connect the aims and plans, the steps of the project, the techniques and technologies involved, the disagreements, refinements, and discarded routes that lead to its concretization. They follow and track the process of construction, the materials used, the companies involved in the procurement, the actors assembling the parts, laying out bricks, piecing together the roof, and extending wires. They trail and record who delivers the building to whom, how the building is occupied, how it is used, modified, repaired, maintained, resisted, and narrated. The assumption is that mapping the complex dynamic emergence of a building involves proliferating actors, texts, entities, and relations, and reaching beyond the usual human-centric agencies, thereby pluralizing and making posthuman the perspectives on design and architecture (Yaneva, 2022).

This approach has been described as *mapping controversies*. Controversies mark points of disagreement in the design and building process that bring together all kinds of human and non-human actors (Yaneva, 2012, 2022). Controversies reveal how design is always in a state of fluidity, where attending to problems, trials, modifications, and new associations has the potential to generate a finer picture of what is involved. This method of analysis has been repeatedly applied to the study of civic buildings and buildings of science, but rarely to school buildings (Backwell & Yaneva, in press). Our project extends this Terrestrial approach by mapping the controversy of post-pandemic school design, following the students and staff as they re-inhabit, re-shape, and re-imagine school buildings after Covid-19.

Many contemporary architects and designers are pursuing posthuman and Terrestrial projects of creative mapping and visualization showing complex networks of distributed agency, influenced by Latour: Design Earth (Ghosn & Jazairy, 2019) Lateral Office and Zurkow (Dwyre et al., 2022), Studio Tomás Saraceno (Enderby, 2022), Bryan Cantley (Baldwin, 2023), Sarah Sze (Latour, 2020), and others. For instance, in *Terraforma: A Book of Speculative Maps*, Aït-Touati et al. (2022) discuss the importance of inventing visualizations that follow the trajectories of those living in/on the ground so as to render visible and knowable how such movements shape space through collaborative and conflicting trajectories and mobilities. They transform the conventional perspectives of grid-based maps and allow viewers to re-imagine the position of human and non-human agencies in relation to Earth's strata, the atmosphere, deep time, the future, and so on. We have been captivated by how *Terraforma's* techniques of visualizing space (maps, drawings, models, installations) achieve a transdisciplinary image of complex multi-agent political ecologies. Their novel visualizations disrupt the perceptual habits instituted by modern ideas of space and matter, presenting more troubled, hybrid, and partial worlds. They function as a means for reckoning with the sensuous, affective, and cognitive dissonances that such worlds provoke, offering their viewers potential ways of connecting their individual concerns to larger collective entanglements and responsibilities (see also Ghosn & Jazairy, 2019).

Consider *Map VII (Re)Collection* (Figure 9.1), where the ruins of industrialization form "a new topography" (Aït-Touati, et al., 2022, p. 173) of situated wastelands, each with different "degrees of habitability" (p. 179). The map merges anthropogenic chronology and speculative futures, visualizing an exploding earth, coded ironically according to a legend of "erasure, rehabilitation, extraction, reinterpretation, regeneration." Such spatial plotting dwells on the im/possibility to "rehabilitate" and "recollect" such ruins. The map depicts paradoxical potential scenarios simultaneously, splintering the very idea of a single inert earth awaiting planned human salvation. And yet the very gesture of the map makes visible and thinkable what is invisible in most maps, a ground that is differently active after the devastating transformations of anthropogenic activity. Resonantly, Aït-Touati et al. (2022) describe their maps as "living maps" rather than "maps of the living," "not only because [living maps] are always under construction, always moving,

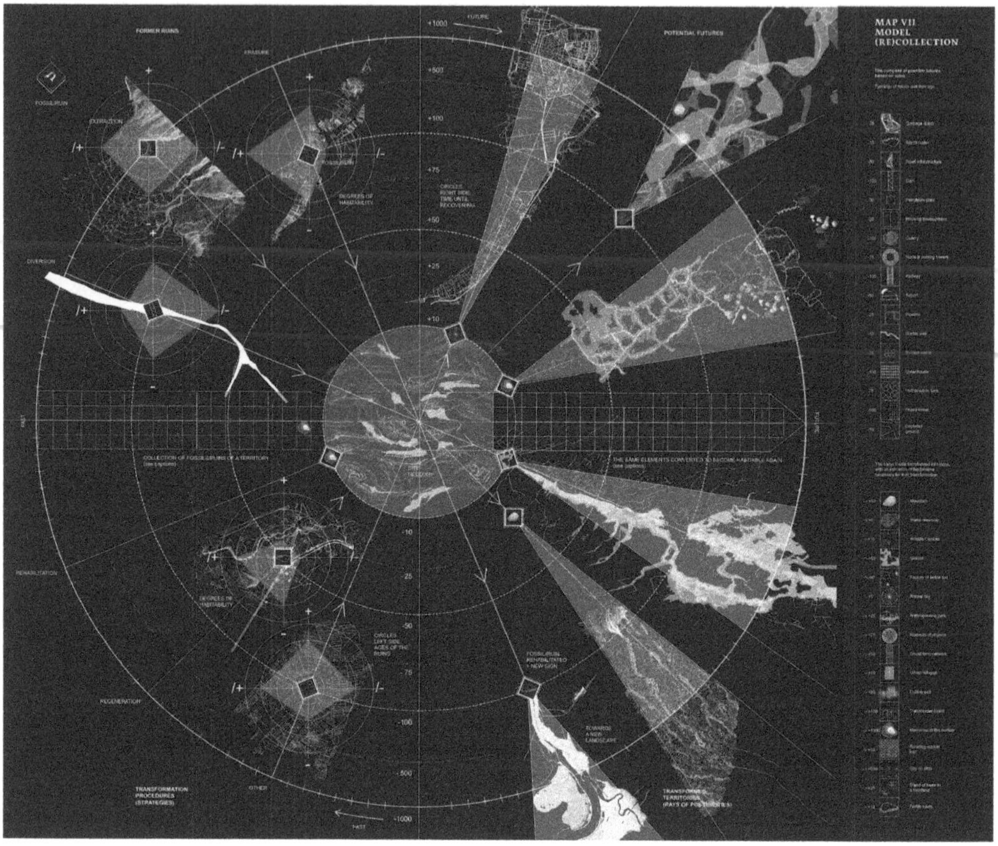

FIGURE 9.1 Map. VII (Re)Collection. Aït-Touati et al. (2022), Terraforma: A Book of Speculative Maps (pp. 182–183). MIT.

Source: Reproduced with permission of Alexandra Arènes.

but because they try to capture and integrate the part of production that is carried out by living things" (p. 19; see also, Knight, 2021). In our school mapping project, discussed below, we take up and use this idea of living maps.

Exploring, sensing, and visualizing the school building as a Terrestrial space

Our ethnographic project applied these design principles and cartographic experiments, extending their relevance beyond the process of architectural planning, to focus on the continued life of post-pandemic school buildings. The ethnographic work was organized across two phases, each five months in duration. In phase one, we interviewed staff and developed workshops and discussions with a group of 14 students (16-year-olds). We used school blueprints during the interviews as well as walking tours to begin to understand how participants connected with specific spaces and conventional maps of the building. The workshops involved experimental situations with sensor technologies and an engagement with spatial problematics like those of perspective, atmosphere and envelop, trajectory and movement, threshold and flow, and sound and air. The data generated included video, sound, photographs, artefacts, and text. Phase two continued with workshop activities, creating new data during various futuring practices, which sometimes included images and models derived from phase one data.

A central aim of the workshops was to transform and enrich students' spatial imagination through experiments that intensified embodied action and sensation in the school building. Due to our focus on the *Terrestrial*, we invested in experiments that pushed spatial experience beyond anthropocentric frames and opened more imaginative ways of inhabiting complex environments (Latour, 2005). More specifically, we were interested in the multiplicity of emotions and sensations fuelling the atmosphere in the post-pandemic school building. Many of these circulate under the threshold of human perceptibility (Fuller & Weizman, 2021) and were pre-conscious or pre-personal (Massumi, 2002). Our aim was to design experiments in sensing the architecture so that the building atmosphere might become visible and tangible in concrete visual maps and models. We used sensor technologies (cameras, air sensors, sound recorders, iPads) and software (Procreate, Flow, Polycam, Hyperspektiv, Outdooractive, Quik, Autocad, Reshoot 360) to move with students as they explored familiar spaces. This technology and software were especially fruitful for "sensing" the different kinds of agencies distributed across the school building. The ethnography rendered the more-than-human relationality visible but also highlighted the provisionality of the built environment and current spatio-temporal compositions, revealing new aspects of transindividual life in the building.

In one of the initial workshops, we explored the problematic of perspective and architecture. We experimented with a GoPro Max camera and the software Reshoot 360 for iPad 2020 to generate 360 images of school spaces chosen by the students. Playing with the combination of camera and software, students were able to bend and distort the rectilinear walls, floors, desks, and seats, thereby recomposing the space according to circular, concave, and convex geometries. In another workshop, we grappled with the problematic of atmosphere and envelope. Students used LiDAR scanners to feel the interior skin of the building, mapping the architectural envelop (Carpo, 2017). Scanning the space as a 3D volume demanded that students move slowly and carefully along walls and into corners, and transformed key rooms into glitchy digital boxes that could be examined as 3D models on the iPads, using Polycam software (see de Freitas & Trafí-Prats, 2023). The resulting ragged, imperfect but also hyperrealist 3D models generated with the app Polycam raised students' curiosity. Students talked about the queer perspectives on the *doll house* 3D images afforded through jagged openings in the digital model. This also triggered discussion about their own hyper-visibility in the school and the absence of spaces where they might not be seen by others. This part of the digital ethnography provoked some students to imagine ways to escape (LiDAR) vision (see chapter 8 for more on this).

In another workshop, we explored the problematic of trajectory and flow in the building. For this, we used some images of spaces that we had previously solarized in Photoshop to make them appear as line sketches. We uploaded these images to the app Procreate, which is widely used for digital painting and illustration, and invited students to trace their daily itineraries. In other workshops, we explored flow and daily trajectories, both to the school and within the school. Students drew their daily journey to school on blank maps of the region. We taught the students some basic features of the Procreate app and how to use the Apple Pencil so they could draw various lines over a given image or map of the school space, depicting the movement of bodies. We showed them how to record and compile the layers of drawn lines into a video file, showing how trajectories unfolded and overlapped across the map over time. These trajectory maps were gestural and tentative, capturing their spontaneous hand movements as they simultaneously observed the space and recorded their observations. In all the fieldwork, students acted as privileged informants and data creators, who revealed key aspects of their lived experience.

A series of workshops investigated the school building using artist and designer Ling Tan's SUPERPOWER app[1] installed in students' phones. The app allowed students to map spaces in the building by integrating a rating scale of good-bad spaces, photographs, texts, and geolocation related to the school floorplan. The collectively generated data revealed six extremely important spaces, associated with concerns and intense emotions – the sixth form hub,[2] the sixth form study room, the main stairs, the atrium, and the atrium's void. Further to this, Tan led the students in creating conceptual cardboard prototypes that imaginatively responded to the collected data and incorporated possible improvements. Many of the prototypes, and more especially those concerned with the transformation of the sixth form hub, the study room, and the atrium's void, demonstrated a repeated interest in creating interior spaces to intensify small group sociality, using things like circular sitting pods, higher walls, curtains, and stand-alone pavilions. Other prototypes altered the architecture for new kinds of mobility. For instance, one of the prototypes proposed to add a slide to the main stairs "for moments of the day when one wants to let go." Another one incorporated some trampolines in the sixth form hub as a way of having "something playful," yet regretting that "they won't be allowed because of safety." Some designs made explicit how dwelling in certain spaces was debilitating, due to the hardness of the furniture and "with no space where to be and talk," referring not only to the lack of space to hang out, but to the impossibility of hearing each other in the open atrium where noise had become a significant school problem. Different designs noted the "flatness" and "dullness" of spaces in the building, suggesting the use of plants in the atrium, the painting of some walls in warmer colours, and the opening of windows in the study room. The cardboard prototypes helped to show how the students' relationship with the building involved a complex mixture of affective forces and pragmatic constraints pertaining to mobility, visibility, and atmosphere.[3]

Living maps

We discuss in detail here *The School Building Living Map* (Figure 9.2), which was designed to incorporate diverse datasets, including interview data with teachers and staff, discussion data from student workshops, and other observation field data during the ethnography. Compiling this data into one visualization was incredibly challenging and involved using many of the creative mapping techniques developed by Aït-Touati et al (2022), and more particularly in the "Living Landscapes Model" (pp. 98–119). The map utilizes a visual token from the BDP blueprints – that being the bracket shape – and layers various kinds of coded behaviour onto it. The four brackets demarcate each of the four floors, which are colour-coded to match the differently coloured walls of each floor. The map also includes the outline of the main elements built around the open atrium space (theatre, offices, main stairs). The map makes visible a palimpsest of accumulated layers of activity organized in temporal refrains that were seen to repeat daily. Different from the standard architectural cross-sectional drawing, which separates activities, assigning each to a particular floor, the map shows all four floors overlayed, superimposing multiple layers of activity in ways that are visually challenging, yet actual. This alternative approach was chosen to better show how the open space atrium made these different activities visible and open to surveillance, while also acting as the key void in the interior of the school, a space for imagining escape from containment, as well as different kinds of mobilities.

After reviewing all the data, we identified 12 ephemeral behaviours and spatial practices that together produced the living space in the course of any one day. Each behaviour is locational and tagged to a particular part of the school, as well as being associated with specific agents. Following Aït-Touati et al. (2022), we characterized these social practices in terms of physical activities that are used by animals when negotiating an environment. The behaviours are meant to evoke the kinds of embodied activities that might be represented in a description of a complex ecology where different species interact, each habituated to its own activity. These are: *flocking* where at the beginning and end of the day staff meets and socializes around the large tables in the atrium which act as a fixing agent; *burrowing* where groups of students nestle in corner spaces outside the classroom using them as temporary refuge; *cascading* where large numbers of students leave the classrooms and flood the atrium; *shuffling* where students and staff travel to the next activity at the end of a period; *corralling*, where a group of students gathers and takes control of a parcel of space, like the sixth form hub; *searching*, where the facilities staff scatter through areas of the school as maintenance is needed; *sweeping*, where the cleaning staff moves linearly and continuously from the lower to the upper floors locating and removing dirt; *gliding*, where groups of seagulls fly outside waiting to collect student food scraps; *cawing*, where various teachers raise their voices to compete for attention in open classroom spaces; *resource guarding*, where staff from the finance department remain in place to guard the school's safe; *sentinelling*, where staff use vantage points in the open atrium to remain in a constant state of vigilance; and *tracking*, where teachers use architectural features like semi-open doors to track student behaviour inside and outside the classroom.

The School Building Living Map presents a map that is descriptive, inscriptive, and generative (Dortdivaniloglu, 2018). The map is descriptive and inscriptive in the sense that it displays a concrete organization of actants and their activities, attending to daily normative habits that we can call "inscriptive." The map is a diagram of how these 12 kinds of behaviour constituted a living landscape that materializes practices of control, escape, refuge, delay, crowding, flow, vanishing, dissipation, etc. The map is also generative because the process of layering and connecting the 12 behaviours effectively produces a landscape with depth and complexity, a multiple-school much like Latour's multiple-earth, co-produced through behaviours that are sometimes at the level of governance, while more often are at the level of minor gestures. These behaviours have been concatenated in the map into six explicitations that show how behaviours compose with other behaviours at different moments of the day. These six groupings are offered as examples of how particular kinds of temporal junctures – the end of class, for example – trigger a set of activities, which bring different kinds of actants together. The living map is temporal in showing how the life of the building unfolds in daily itineraries and activities, which are brought together at particular moments in time. The map also materializes a Terrestrial multiple-school, revealing the micro-territories that emerge as part of the political

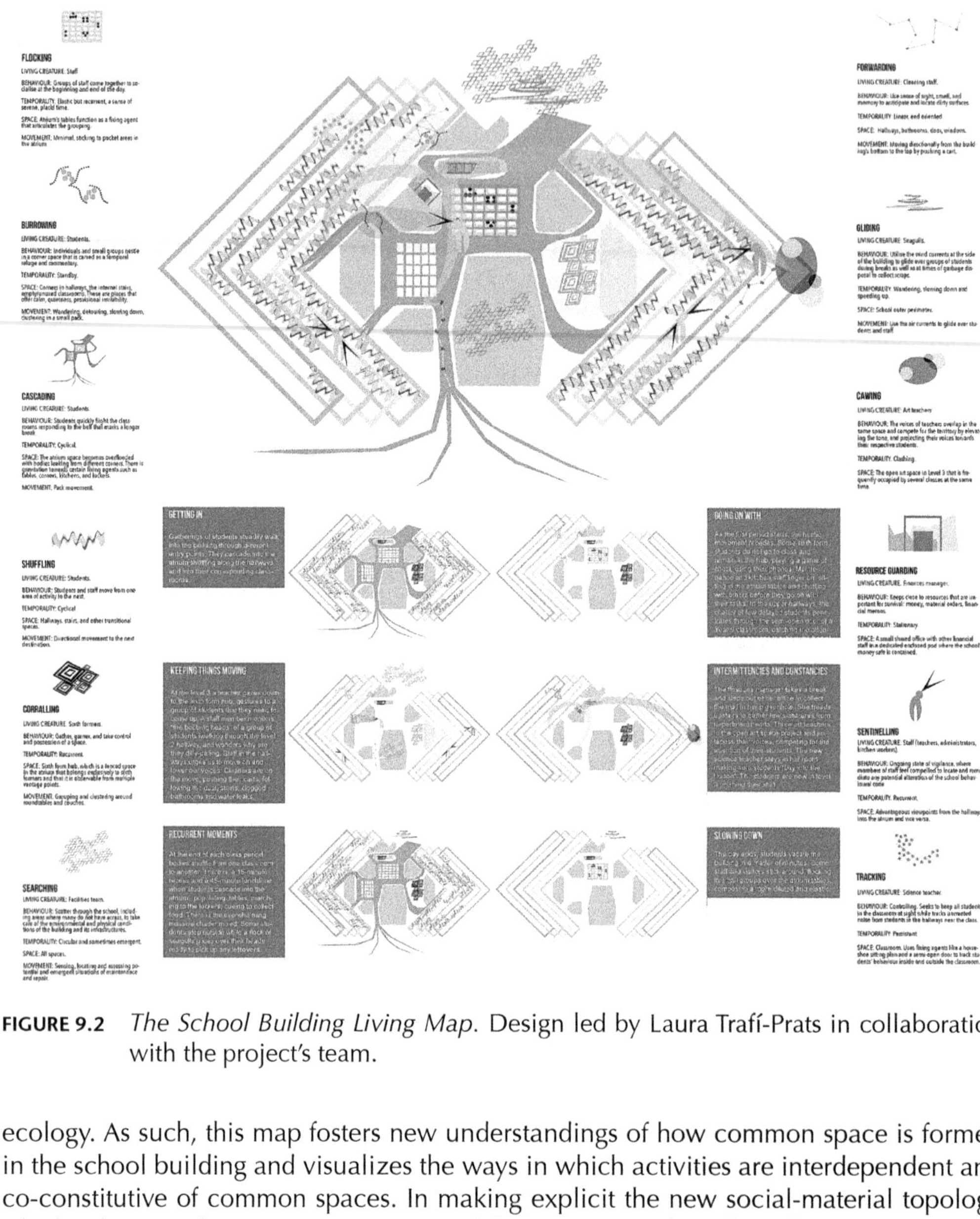

FIGURE 9.2 *The School Building Living Map.* Design led by Laura Trafí-Prats in collaboration with the project's team.

ecology. As such, this map fosters new understandings of how common space is formed in the school building and visualizes the ways in which activities are interdependent and co-constitutive of common spaces. In making explicit the new social-material topology of school space, the map exposes some of the tensions and conflicts that emerge among different agents in the overlapping layers. It reveals the ways that certain behaviours are convergent, while others are divergent and contribute to an agitated atmosphere.

Speculative models using 3D software

After identifying salient features of student maps, models, and other futuring practices during the workshops, and correlating these to ethnographic and observational data, we were able to modify existing architectural plans and build 3D models of the lived school building using CAD software. We exploited the possibilities that CAD software offers to set any digital model into continued variation (Carpo, 2017), allowing us to modify the models as we continued to review our data, while offering an important mode of spatial analysis that attends to the relationality of the built environment and therefore the ways in which our students were implicated in a "whole" space of interconnected agents (Ghosn & Jazairy, 2019). These five models synthesized student data on affect and spatial sense, and appealed to the workshop futuring practices and speculative mapping and modelling techniques we cited above. We named the five models: *The Archimedea* (Figure 9.3), *The Wormhole* (Figure 9.4), *The Seagull's Swooping Circular Path* (Figure 9.5), *The Draped Lily Pads* (Figure 9.6), and *The Bubble* (Figure 9.7).

The Archimedea consists of stacked polyhedral forms (*Rhombicuboctahedra)* which seem to grow with torsion and detach as they rise up in the atrium, from the ground floor to the second floor (Figure 9.3). This strange aggregate form resembles a complex molecule, perhaps composed of viral particles chemically fused or linked, some transparent and ghost like, offering a viral passage between key locations in the school. This model involves individuated pods assembled together into a precarious discontinuous structure, much like the student body as it re-assembled itself post-pandemic. The corridors and open spaces appear littered with the polyhedra, cluttering and filling the space, functioning both as a new mode of occupation but also as obstacles countervailing previous forms of mobility. The model reflects the post-Covid environment in which this research occurred, where the mobility of the students was radically reconceptualized in terms of unusual highly scripted navigations, evoking new postures and movements such as hunching, avoiding, and masking. This model also speaks to students' wish for other geometries and structures, and emphasises the play of Euclidean constraint on the environment. As such, the design poses multiple questions about students' sense of their co-habitation and connectivity in this post-pandemic space.

The *wormhole* model explores ideas of meandering, indirect or inefficient movements. In this case, the transit from the sixth form hub (atrium) up to the platform (second floor) is through a stretchy see-through netted tunnel that twists across the atrium's void (Figure 9.4). The title of wormhole is in reference to an astrophysics concept that refers to links between two different points in space-time, creating a shortcut that could reduce

FIGURE 9.3 *The Archimedea.* Images generated with AutoCAD (for 3D drawing) and Lumion (for rendering). Design led by Izzy McCauley in collaboration with the project's team.

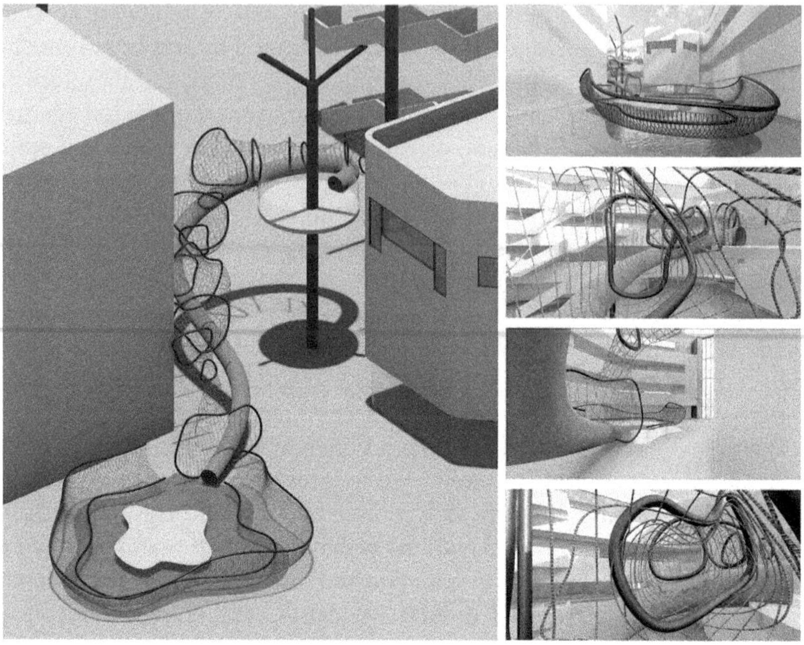

FIGURE 9.4 *The wormhole*. Images generated with AutoCAD (for 3D drawing) and Lumion (for rendering). Design led by Izzy McCauley in collaboration with the project's team.

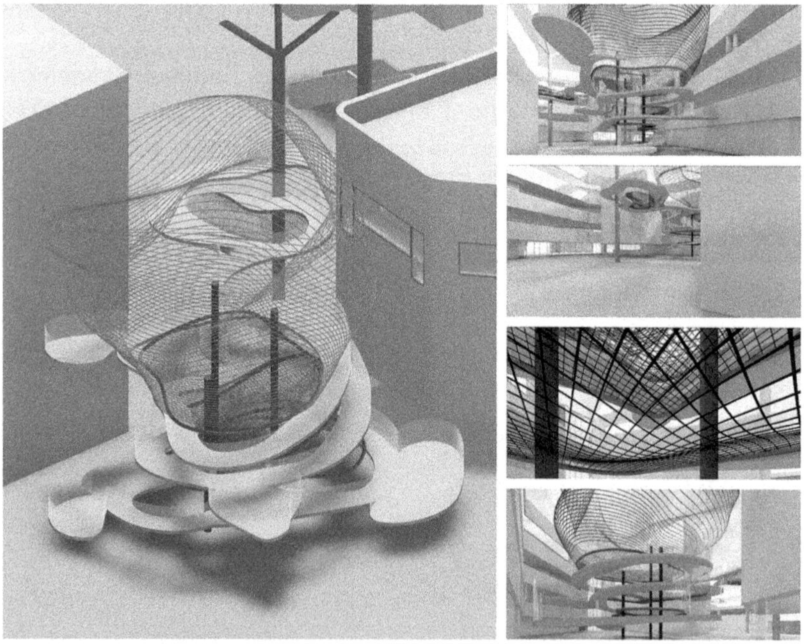

FIGURE 9.5 *The seagull's swooping circular path*. Images generated with AutoCAD (for 3D drawing) and Lumion (for rendering). Design led by Izzy McCAuley in collaboration with the project's team.

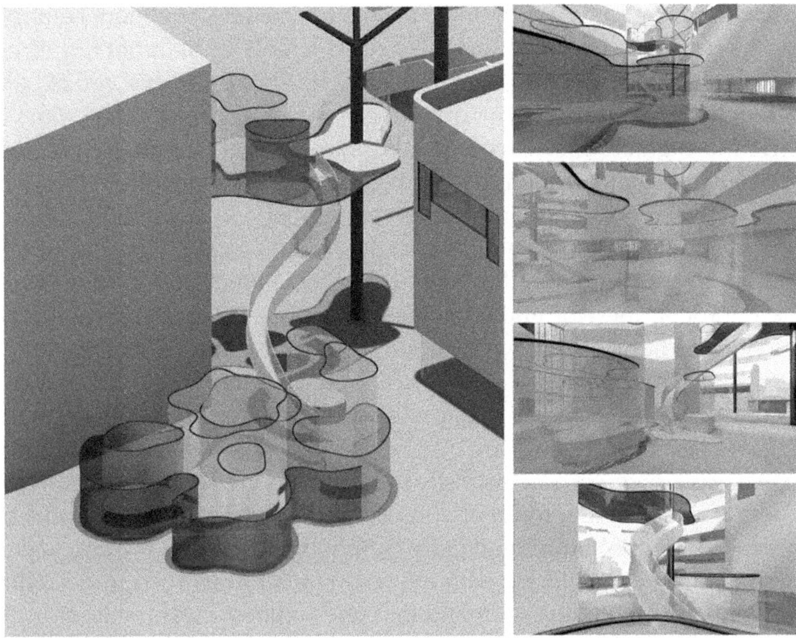

FIGURE 9.6 *The draped lily pads*. Images generated with AutoCAD (for 3D drawing) and Lumion (for rendering). Design led by Izzy McCauley in collaboration with the project's team.

FIGURE 9.7 *The bubble*. Images generated with AutoCAD (for 3D drawing) and Lumion (for rendering). Design led by Izzy McCAuley in collaboration with the project's team.

travel time and distance. The concept is found in science fiction, where travellers can move from one world to another, without having to follow the usual rules of motion, but also in solutions to Einstein's theory of general relativity. The wormhole speaks to the students wish to be in their two preferred spaces without having to traverse territories that feel dangerous. If the students can travel in these wormholes, they can avoid the exposed staircases that flank the atrium and are part of the panoptic architecture of the school building. They can occupy these two spaces linked by positive affect, and move back and forth, without risking exposure to the forces that cause them anxiety. The model embodies these ideas in various ways, as the tunnel narrows considerably at different points and makes passage difficult although not impossible because the fibre is flexible and expands. Evoking a science-fiction imaginary, the wormhole is a way of escaping space-time trappings, and traveling into higher-dimensional spaces, introducing a dramatic change in location, and the journey is felt through the vibration of the net, rather than through the eyes and their perspectival tendencies (and associated panopticon capacities), as bodies move from one end to another. Running adjacent to the netted tunnel, there is a narrow tubular slide made of rigid, opaque, smooth material. The movement on the slide contrasts with the movement through the wormhole because it operates through conventional Newtonian mass, so that bodies speed up with gravity as they descend, reflecting students desire for an embodied feel of exhilaration and "letting go." The model resonates with the work of Studio Tomás Saraceno and their exploration of a politics of space through the making of alternative webbed models of transit and communication (Galantino & Coccia, 2020).

The seagull's swooping circular path also reveals alternative mobilities in the school. It is composed of a rambling ramp that conjures the swooping circular paths of the seagulls flying near the school building (Figure 9.5). Here the nets are more expansive and allow for nesting or perching, along with the resting platforms. The ramp is entirely disorienting and continues the focus on inefficient and indirect movement and complicates the visual perspectives on the surrounding the atrium. Additionally, the ramp slopes and angles inwards, making movement precarious but exciting, and challenging balance. The net also requires crawling, jumping and bouncing, decentring the image of bipedal man and grid-controlled movements. It plays too with possibilities of drifting and being suspended in the air.

The draped lily pads are made of curtained spaces that can be opened and closed. Curtains were proposed by the students in their prototypes and mappings, to create smaller private spaces within the shared space of the atrium (Figure 9.6). The drapery hangs from mechanical rails and has different opacities creating a fluid, soft, and metamorphosing perimeter. The name "lily pads" refers to the students' interest in draped areas surrounded by moats of water, here visualized as rings of water that isolate, protect, and delimit the sixth form hub from other atrium occupants and passers-by. Both curtains and moats offer radically different textures (soft, velvety, moist, fluid) to the school building, and change the haptic nature of its surfaces. Students spoke of how the sounds of the curtains opening and closing and the water moving would increase the affective-sensory nature of the model, demarcating the area through its own soundscape. Like other models, these lily pads offered an alternative within a space that addressed students in terms of being "ready to learn" and a "school citizen," offering a dreamlike space where they might "take a break," "relax," "let go," "hang out with friends," "be by yourself."

The bubble model incorporates translucent and transparent bubbles of inflated plastic, amplifying the fantastical references to waterfalls, steppingstones, and grass in student data (Figure 9.7). These fantastical references came forth in the workshops, alongside the expressions of anxiety, precarity, vulnerability and awareness of interdependence. *The Bubble* reveals a controlled sociality of shared air and breath. Through constant curvature, the bubble shape shows that the school was a space of containment and captivation, as a kind of lockdown during the pandemic. The open air of the atrium is entirely rejected in this model, representing the force of government regulatory interventions to moderate and manage toxic air quality, under spherical containment.

The five models draw from SF imaginaries as a way of thinking with the limitations and possibilities of school space. They can function as a springboard for discussing the ways in which learning environments figure in the lives of students, and how the school building is "more contingent than it is presented to be" and thus the models "demonstrated and proved that the world in which we live is neither necessary nor complete" (Lecompte, 2020, p. 168). Indeed, they make visible experiences of the school building where space is neither a fixed geometry, nor the background in which bodies undertake actions, but is carried, moved, and extended through inhabitation and improvisation (Enderby, 2022). These are visualizations that allow viewers to imagine the perils of navigating the current environment, and possible ways of intervening in the school building to address its affective shape (Ghosn & Jazairy, 2019). They perform a "tragicomic perspective" (Anker & Anker, 2019, p. 210) in which actual accounts about noteworthy social, affective, and environmental issues affecting life in the building repeatedly evoked by the students are addressed with creativity, humour, and the imagination. We thus approach school architectures *from within* their complex unfolding (Latour, 2018), making explicit various implicit structures that were not visible prior to the research, revealing the fluid space between what it is and what is possible (Halse & Boffi, 2016), so that others can discuss local adjustments that support their own scenarios of how and why to imagine and create other school architectures.

Conclusions

Our maps and models demonstrate how the complex social-material ecology of post-pandemic school space emerges through various modes of real and imagined engagement. Our focus has been on the emotional impact of the pandemic, as it implicitly and explicitly shaped fears and desires about proximity, mobility, and inhabitation. Students' spatial practices underwent radical transformations; their schools became increasingly about surveillance, their bodies treated as viral vectors. These speculative renderings of our data explicitly avoid resolving multiple disjunctions and tensions, letting the data with its anxieties, sensitivities, and disparate modes of attention come to the fore. As such, the maps and models show the differentiated *commons* of school space. Latour (2005) argues that in such a heterogenous political ecology the common is not given but needs to be composed. He adds that such composition is only possible if we become "interested in differences that seek what they have in common" (p. 220). Following this ethos, our collaborative digital-sensory ethnography worked with the idea that there is no common "without practices that entangle people" (Stengers, 2023, p. 179). But we also understood that such *commoning* already involved both the technical and the more than human. We showed students how to make visible the alien sensitivities of sensor technologies and software to conjure the pre-personal, atmospheric, affective, and animal dimension of spatial life in schools. We have called these *futuring practices* because they intensify the potential for re-imagining and re-shaping the school building.

In alignment with the work of contemporary architects, artists, and designers, we have argued that speculative spatial methods allow us to share research findings in visual models. The data has been "written up" in maps and models. This entails describing space anew as a complex, multiple, open-ended, Terrestrial actor, while imagining it beyond the tropes of realism, optimization, and determinism that habitually frame the discourse on school buildings. Thus, *the School Building Living Map* connected locational behaviours of different actants and overlayed them to visualize daily space-time refrains. We pursued an alternative representation of the building that mattered the space differently, drawing attention to micro-territories, interdependencies, tensions, and various forms of common life that emerged. Similarly, the CAD software allowed us to transform the designs so that they might better evoke the spatial relational ecology, as we continued reviewing our data and discussed the ideas with the students. Both processes effectively foregrounded the potential of speculative thinking, cultivating an approach to architecture and the built environment that was *responsibly* future-oriented.

Finally, this project has raised new questions about how architecture might become more responsive to our changing social-material conditions, as ecological crises and issues of spatial injustice proliferate. How might these kinds of speculative experiments help us formulate new modes of earthbound inhabitation and change our spatial practices and modes of belonging? In such imagined futures, who will be able to participate in these altered earthly compositions? Which sensitivities will be solicited and which will go undeveloped? How will researchers attend adequately to the emotional, cognitive, and affective dissonances proliferating in these learning ecologies? Following Latour, how might we support a more inclusive relational ecology, honouring and engaging with the metamorphic Terrestrial through technical, speculative, and responsible modes of commoning?

Notes

1. Ling Tan's website features few of the creative projects that she has developed using the SUPERPOWER app, https://lingql.com/superpower-romania/.
2. In the UK sixth form corresponds to the last two years of high school, with students aged 16–18.
3. Documentations of these workshops can be found in the project's website: www.schoolatmospheres.net.

References

Aït-Touati, F., Arènes, A., & Grégoire, A. (2022). *Terra Forma: A book of speculative maps*. (Trans. Amanda DeMarco). Cambridge: MIT Press.

Anker, P., & Anker, N. E. (2019). Anthropocene architecture: Design Earth's *Geostories*. In R. Ghosn. & E. H. Jazairy (Eds.), *Geostories: Another architecture for the environment* (pp. 206–213). Barcelona: Actar.

Arènes, A., Latour, B., & Gaillardet, J. (2018). Giving depth to the surface: An exercise in the Gaia-graphy of critical zones. *The Anthropocene Review*, 5(2), 120–135.

Blackwell, B., & Yaneva, A. (2024). School buildings as performative machines: the new architectural devices of control. *Social & Cultural Geography, 25*(9), 1451–1470. https://doi.org/10.1080/14649365.2024.2334947

Baldwin, P. J. (Ed.). (2023). *Speculative coolness: Architecture, media, the real and the virtual.* Routledge.

Burke, C. (2013). *A life in education and architecture: Mary Beaumont Medd*. Farnham: Ashgate.

Burke, C., & Grosvenor, I. (2008). *School.* London: Reaktion Books.

Cardellino, P., & Woolner, P. (2019). Designing for transformation – A case study of open learning spaces and educational change. *Pedagogy, Culture & Society*, 28(3), 1–20.

Carpo, M. (2017). *The second digital turn: Design beyond intelligence*. Cambridge: MIT Press.
Cleveland, B., & Fisher, K. (2014). The evaluation of physical learning environments: A critical review of the literature. *Learning Environments Research*, 17(1), 1–28.
Cooper, I. (1981). The politics of education and architectural design: The instructive example of British primary education. *British Educational Research Journal*, 7(2), 125–136.
Cuban, L. (2004). The open classroom: Were schools without walls just another fad? *Education Next*, 4(2), 68–72.
Daniels, H, Stables, A, Tse, H., & Cox, S. (2019b). *School design matters*. London: Routledge.
Daniels, H., Tse, H., Stables, A., & Cox, S. (2019a). Design as a social practice: The experience of new-build schools. *Cambridge Journal of Education*, 49(2), 215–233.
de Besten, O., Horton, J., Adey, P., & Kraftl, P. (2011). Claiming events of school (re)design: Materializing the promise of *Building Schools for the Future*. *Social & Cultural Geography*, 12(1), 9–26.
de Freitas, E., & Trafí-Prats, L. (2023). Atmospheric data and software arts: New ways of investigating the built environment. In A. Mazzei & A. Y. Jackson (Eds.), *Postfoundational approaches to qualitative inquiry* (pp. 51–64). New York: Routledge.
Deed, C., Lesko, T. M., & Lovejoy, V. (2014). Teacher adaptation to personalized learning spaces. *Teacher Development*, 18(3), 369–383.
Dortdivanlioglu, H. (2018, September). The diagram in continuum: From inscription to generation of form in architecture [Conference paper] ARCC Conference Repository. https://www.arcc-repository.org
Dovey, K., & Fisher, K. (2014). Designing for adaptation: The school as socio-spatial assemblage. *The Journal of Architecture*, 19(1), 43–63.
Drucker, J. (2021). Visualization. In N. B. Thylstrup, D. Agostinho, A. Ring, C. D'Ignazio, & K. Veel (Eds.), *Uncertain archives: Critical keywords for big data* (pp. 561–568). Cambridge: MIT Press.
Dwyre, C., Perry, C., Salomon, D., & Velikov, K. (2022). *Ambiguous territory: Architecture, landscape, and the postnatural*. Barcelona: Actar.
Enderby, E. (2022). Floating on an ocean of air. In E. Enderby, Studio Tomas Saraceno & The Shed (Eds.), *Tomás Saraceno: Particular matter(s)* (pp. 9–34). Verlag der Buchhandlung Walther und Franz König.
Fuller, M., & Weizman, E. (2021). *Investigative aesthetics: Conflicts and commons in the politics of truth*. New York: Verso.
Galantino A., & Coccia, E. (2020). *Tomás Saraceno: Aria*. Venice: Marsilio.
Ghosn, R., & Jazairy, E. H. (2019). Another architecture for the environment. In R. Ghosn & E. H. Jazairy (Eds.), *Geostories: Another architecture for the environment* (pp. 10–27). Barcelona: Actar.
Halse, J., & Boffi, L. (2016). Design interventions as a form of inquiry. In R. C. Smith, K. Tang Vangkilde, M. C. Kjærsgaard, T. Otto, J. Halse, & T. Binder (Eds.), *Design anthropological futures* (pp. 89–103). London: Bloomsbury.
Haraway, D. (2013). SF: Science fiction, speculative fabulation, string figures, so far. *Ada: A Journal of Gender, New Media, and Technology*, 3, 1–18. Available at https://scholarsbank.uoregon.edu/xmlui/handle/1794/26308.
Harris, D., & Rousell, D. (2022). Posthuman creativities: Pluralist ecologies and the question of how. *Qualitative Inquiry*, 28(5), 427–434.
Kay, A. D., Scherber, E., Gaitan, H., & Lovelee, A. (2019). Transitional ecology: Embedding ecological experiments in temporary urban public art. *Journal of Urban Ecology*, 5(1), 1–7.
Knight, L. (2021). *Inefficient mapping: A protocol for attuning to phenomena*. Santa Barbara: Punctum.
Kraftl, P. (2012). Utopian promise or burdensome responsibility? A critical analysis of the UK governments' building schools for the future policy. *Antipode*, 44(3), 847–870.
Kraftl, P., McKenzie, M., Gulson, K., Accioly, I., Blackmore, J., Burke, C., Perrota, C., Clarke, D. A. G., Daniels, H., Fregoso Bailon, R. O., Goodyear, V., Goodyear, P., Gunasekara, I., Hartong, S., Hickman Dunne, J., Howard. S., Lupinacci, J., Mcphie, J., & Wood, A. (2022). Learning spaces:

Built, natural and digital considerations for learning and learners. In S. Bugden & G. Borst (Eds.), *Education and the learning experience in reimagining education: International science and evidence based education assessment* (pp. 450–526). UNESCO MGIEP. Available at https://mgiep.unesco.org/iseeareport.

Latour, B. (1993). *We have never been modern*. Harvard: Harvard University Press.

Latour, B. (1997). Trains of thought: Piaget, formalism, and the fifth dimension. *Common Knowledge*, 6, 170–170.

Latour, B. (2005). There is no terrestrial world. In M. Ohanian & J.-L. Royoux (Eds.), *Cosmograms* (pp. 211–222). London: Lukas & Sternberg.

Latour, B. (2008, September, 3). *A cautious Prometheus? A few steps towards a philosophy of design (with special attention to Peter Sloterdijk)* [Conference Presentation]. Networks of Design Meeting of Design History Society Falmouth, Cornwall, United Kingdom.

Latour, B. (2011, November 7). Waiting for Gaia. Composing the common world through arts and politics. Available at https://www.bruno-latour.fr/sites/default/files/124-GAIA-LONDON-SPEAP_0.pdf.

Latour, B. (2017). *Facing Gaia: Eight lectures on the new climate regime*. Cambridge: Polity.

Latour, B. (2018). *Down to Earth: Politics in the new climate regime*. Cambridge: Polity.

Latour, B. (2020). Sara Sze as the sculptor of critical zones. In B. Latour & P. Weibel (Eds.), *Critical zones: The science and politics of landing on Earth* (pp. 160–163). Cambridge: MIT Press.

Latour, B. (2022). Foreword: How to fight about space. In F. Aït-Touati, A. Arènes, & A. Grégoire (Eds.), *Terraforma: A book of speculative maps* (pp. 4–8). Cambridge: MIT Press.

Latour, B., & Weibel, P. (2020). Seven objections against landing on earth. In B. Latour & P. Weibel (Eds.), *Critical zones: The science and politics of landing on Earth* (pp. 12–19). Cambridge: MIT Press.

Lecompte, J. (2020). Can the possible exist in physical form? In A. Beech & R. Mackay (Eds.), *Construction site for possible worlds* (pp. 145–170). New York: Urbanomic Media LTD.

Massumi, B. (2002). *Parables for the virtual: Movement, affect, sensation*. Durham, NC: Duke University Press.

Niemi, K., Minkkinem, J., & Poikkeus, A.-M. (2022). Opening up learning environments: Liking school among students in reformed learning spaces. *Educational Review.* https://doi.org/10.1080/00131911.2022.2098927.

OECD (2017). *OECD framework for a module on the physical learning environment* – Revised edition. Directorate for Education and Skills Education Policy Committee. Available at https://www.oecd.org/education/OECD-FRAMEWORK-FOR-A-MODULE-ON-THE-PHYSICAL-LEARNING-ENVIRONMENT.pdf.

Salmarsh, S., Chapman, A., Campbell, M., & Drew, C. (2015). Putting "structure within the space" Spatiality un/responsive pedagogic practices in open-plan learning environments. *Educational Review*, 67(3), 315–327.

Siguršardóttir, A. K., & Hjatarson, T. (2016). The idea and reality of an innovative school: From inventive design to established practice in a new school building. *Improving Schools*, 19(1), 62–79.

Stengers, I. (2023). *Making sense in common: A reading of Whitehead in times of collapse*. University of Minnesota.

Thorham, H., & Myers, C. A. (2012). Architectures of youth: Visibility, agency and the technological imaginings of young people. *Social & Cultural Geography*, 13(7), 783–800.

Tresch, J. (2005). Cosmogram. In M. Ohanian & J.-L. Royoux (Eds.), *Cosmograms* (pp. 67–76). London: Lukas & Sternberg.

Tresch, J. (2020). Around the pluriverse in eight objects: Cosmograms for the critical zone. In B. Latour & P. Weibel (Eds.), *Critical zones: The science and politics of landing on Earth* (pp. 58–69). Cambridge: MIT Press.

Yaneva, A. (2012). *Mapping controversies in architecture*. London: Routledge.

Yaneva, A. (2022). *Latour for architects*. London: Routledge.

10

NEW EMPIRICISMS IN THE ANTHROPOCENE

Thinking with speculative fiction about science and social inquiry

Elizabeth de Freitas and Sarah E. Truman

Introduction

Much has recently been written about the need to rethink the nature and purpose of social inquiry (Clough, 2009; Maclure, 2013; Ruppert et al., 2013). We hear calls for new empiricisms and demands for speculative and transdisciplinary methods that might break up rigid and ossified research practices in the social sciences (St. Pierre, 2016). Many of these calls share a concern that past theories of sociality have dismissed the vibrancy and animacy of the non-human material world. To that end, new materialist thinkers such as Diana Coole and Samantha Frost (2010) argued that cultural theory must *renew* its commitment to materiality through engagements with 20th century post-classical physical science. Associated attempts to remix the physical and social sciences, in the work of scholars like Manuel Delanda (2015) and John Protevi (2013), have turned to contemporary chemistry and biology, inspired in part by the Deleuze-Guattarian philosophical practice of engaging directly with scientific knowledge. Others have pursued a rapprochement between quantum and social science, showing how quantum theory can be used to make sense of human behaviour (Barad, 2007; Wendt, 2015). Feminist theorists like Elizabeth Wilson (2015) suggest that sociologists would benefit from thinking with new insights from physiological and pharmacological empirical science, while Elizabeth Povinelli (2018) pursues new mixtures of geological and Indigenous knowledges, as a way of reckoning with the pressing problematic of non-life (or geo), in confronting the "end-times" of a particular biosphere. In related efforts, Bruno Latour (2017) and Isabelle Stengers (2018) seek new non-reductive forms of science adequate to the emergent complex naturecultures of Anthropocene[1] life.

Haunting these efforts is the fact that science often acts as a handmaiden to industry and colonial acquisition. The old "white geologies" and biologies encode within them an extractive and enslaving logic, mobilized against racialized people and non-human animals (Yusoff, 2019). Concepts like the Capitalocene (Moore, 2016) and Plantationocene (Tsing, 2017) remind us that the Anthropocene has been manufactured by a portion of humanity invested in accelerated capitalist accumulation and white supremacy. Science

DOI: 10.4324/9781003538769-10

has played a crucial role in shaping this "global" condition as a legacy of European imperialism. And yet it would be foolish to deny scientific knowledge as simply serving the white establishment, particularly today under neoliberal post-truth conditions. Science denialism is on the rise, allied with nationalist anti-establishment movements and libertarian free market interests (Drummond & Fischhoff, 2017). Climate science in particular is contested by segments of the US population as fabricated by scientists with a biased agenda (Lewandowsky et al., 2013). Sociologists and other scientists alike are warning that science is under attack by the Christian right (Alumkal, 2017). The Union of Concerned Scientists and various speakers at the March for Science have declared that President Trump is at war with science. Latour (2017) highlights how the language of "scientific consensus" is precisely what the climate-change deniers pounce on today, identifying themselves with the renegade science of Galileo and his fight against the theocratic establishment.

What new forms of empirical inquiry will emerge in these complex circumstances? As culture and media theorist Kara Keeling (2019) states in *Queer times, black futures*, "Working with and through scientific and technical knowledge in the service of humanistic inquiry is a challenge of world historical significance" (p. 195). Keeling turns to the power of speculative and poetic thought in her work, exploring how speculative thought operates in diverse contexts – in scientific laboratories, artistic studios, financial think tanks, and fiction workshops. In this article, we also turn to speculative thought, and like Keeling we turn to fiction as a way of imagining alternative forms of empirical inquiry. Speculative fiction (SF) often presents cosmic remixings of the socio-material sciences, thereby helping us to imagine a future empiricism for earthbound "terrans" who have forgotten that their planetary dwelling is also a living creature. We follow many others in Science Studies who have taken up SF to think through the many pasts, presents, and futures of science. Donna Haraway (2013), for instance, has played a pivotal role in exploring the role of speculative fabulation in scientific practice. And queer feminist science studies scholar Aimee Bahng (2017) has argued that the science fiction of Octavia Butler (2000) should itself be considered a kind of Science Studies, for how it mobilizes speculative thought to explore radical onto-epistemological modes of alien life, queer notions of reproduction, and decolonial kinship.

In this chapter, we explore the following questions:

1. Are there scientific imaginaries evoked in SF that might be helpful for socio-cultural researchers who are seeking ways to complicate empirical method and methodology?
2. In what ways does SF present alternative research methods for studying pluralist posthuman sympoietic ecologies and new forms of worldly belonging?

We focus on the SF text *Annihilation* by Jeff VanderMeer (2014) precisely because it explores mutation, variation, and destruction, and because the main character is a biologist whose scientific method undergoes transformation in an environmental disaster zone called Area X. We begin to glimpse in Area X both an opportunity and a demand for empirical methods that remix the social and physical sciences to better suit Anthropocene naturecultures. Science in Area X becomes attuned to bio-technical exchange, chemtactile forms of communication, and alt-forms of coupling, mutating, and learning. We analyse the novel throughout this paper, discussing the plot and character development, as a source for new images of empiricism and methodology. In particular, this SF novel attends to the post-truth un/reliability of empirical inquiry in a world where matter-mind mixtures and in/organic variation seem to undermine conventions of scientific observation, classification, and agency.

Thinking with literature

We situate our project in the burgeoning theoretical space of *posthuman ecologies* where various scholars are rethinking the relationship between nature, technics, and human "sense" making (see Braidotti & Hlavajova, 2018).[2] We use the term "ecology" precisely because it has become a paradigmatic concept in Western theorizing of life and society and much else, and because of the way it points to Indigenous and non-Western philosophies of matter-mind relationality and geontology (Povinelli, 2018; Todd, 2016). Additionally, Hörl (2018) describes a "general ecology" that has become "denaturalized" in the Capitalocene, where plastics and digital e-waste are now ecological forces. Morton (2016) calls this "ecology without nature," while Parisi (2017) raises concerns that the new "ecological rationality" is conceived as fundamentally computational. Following these authors, we stretch this term ecology into new territory, attending to the metamorphic nature of matter and trans-species intimacies, while recognizing the non-relational, the separating tendencies, and the "outside" of the ecological paradigm (Neyrat, 2017).

This approach resonates with our previous work on worldly sensibility (see Chapter 6), speculative thought and method (de Freitas, 2017a; Springgay & Truman, 2018; Truman, 2017), and our commitment to exploring posthuman ethnography (see Chapter 5). It continues our work exploring links between qualitative research and new materialisms (de Freitas & Curinga, 2015; Snaza et al., 2016; Truman, 2019a) and probes the fissures between fact and fiction in the shaping of new empiricisms. We have collaboratively developed a method of working with SF texts, exploring the ways in which the story world opens up possibilities for new forms of social-material inquiry (de Freitas, 2017a; de Freitas & Truman, 2019; Truman, 2019b). This involves transdisciplinary "reading" practices, across theory and fiction, in search of alternative images of inquiry. Such work draws extensively from Science Studies, where the anthropological and sociological study of scientific practice has, for decades, shed light on the complexity of empiricisms of all kinds (e.g. Fox-Keller, Haraway, Harding, Longino). Recent journal special issues make the case for an increased take-up of methods from Science Studies in efforts to rethink the intersections between materiality and sociality (de Freitas et al., 2017; de Freitas & Weaver, 2019).

In this chapter, we analyse fiction as a mode of expression that offers insight into future and/or alternative worlds. Our reading of *Annihilation* and other SF texts is meant to offer poignant illustrations that complicate contemporary scientific inquiry and its influence on qualitative research in the Anthropocene. We first discuss SF craft practices and their potential role in expanding our understanding of more-than-human relationality. We argue that SF can make visible the inorganic potentialities and inhuman forces by which a body becomes transindividual and learns to "branch out into territories beyond its own self-maintenance" (Colebrook, 2014, p. 138). It is precisely these kinds of sympathetic relationships that are pursued in particular SF literary texts (often during times of war), when new and different ways of mutating and propagating "offspring" are described, or when trans-species merge and play havoc with conventions of biology, geology, chemistry, genetics, etc. SF narratives allow us to raise important questions in the "end-times" of the Anthropocene, about the future of the sciences and the possibility of generative and reparative sympathies. Our focus on sympathy in this paper, or what Haraway calls sympoiesis – is intended as a way of advocating for a form of inquiry into vast more-than-human meshworks, capable of "hosting" radical incommensurable differences (Chakrabarty, 2014).

Our focus is on how examples of SF pursue an ecological cosmic sympathy with the non-human, and how close readings of these texts allow scholars to think creatively about new kinds of inquiry in the Anthropocene. Many SF texts utilize literary devices involving biochemical metamorphic transitions and altered forms of coupling and reproduction between humans and more-than-human entities (including aliens, bacteria, concepts). We realize that attention to trans-species transits and earthly belonging demands a careful consideration of how the whiteness and misogyny of the SF genre have been (and continue to be) contested through its very possibilities (Bahng, 2017). Thus, while exploring SF for past and possible futures of empirical inquiry, we emphasize the themes of reproduction and mutation in order to *interrogate the biopolitics of organic variation* (Subramaniam, 2014; Willey, 2016) and the *sentimental politics of benevolent sympathy* (Schuller, 2018). In other words, we aim to stay tuned to the complex political ramifications of empirical inquiry in Anthropocene naturecultures.

SF craft

SF has been described as a "literature of cognitive estrangement" (Suvin, 1972, p. 372). This estrangement or defamiliarization occurs when the author builds an imaginative plot that radically alters the "real world" empirical environment, or through language use ruptures taken-for-granted knowledge. Defamiliarization is both a pedagogical tool and a literary device (Truman, 2017). Similarly, new materialist philosopher Rosi Braidotti (2013) uses the term *disidentification* to describe thinking processes that lead to the "…loss of familiar habits of thought and representation in order to pave the way for creative alternatives" (pp. 88–89). In an era of post-truth politics and growing anti-science movements, it might seem like folly to explore cognitive estrangement, defamiliarization, and disidentification with scientific consensus. On the other hand, probing the fissures between fact and fiction might be the very thing we need to do more carefully, under these complex onto-epistemological conditions.

SF stories often raise questions about mutation, metamorphosis, haptic encounters, alien genesis, coupling, reproduction, breeding, continuance, and belonging to worlds in various end-times. Consider, for instance, Afrofuturist Octavia Butler's book *Lilith's Brood*, which introduces the Oankali, an alien species who rescues humans and brings them aboard their ship (which later is found to be a living thing that one can bio-technically interact with) after Earth is rendered uninhabitable by an unnamed war and climate disaster. The Oankali are a species of genetic engineers that evolve through genetic exchange, genetic mutation, and other-species intimacy. During the narrative, Lilith "mates" with an ooili – which is the third gender of the Oankali and serves as a bridge between male and female Oankali and other species they encounter (in Lilith's case, human). After mating with the ooili, Lilith assumes tactile capacities for interacting with the living ship. Lilith's power grows through *sympathetic genetic* exchange with Oankali. She performs a kind of risky diplomacy across conventional divides, a diplomacy that ultimately betrays those conventions, seeking and forming new dependency relations (de Freitas, 2019b).

In African and American author Nnedi Okorafor's *Binti*, the main character is an earth dweller, Binti from the African Himba tribe. She has creative mathematical capacities, and can meld with concepts and formalisms – Binti is a "master harmonizer" of mathematics and can perform a kind of trance-like mathematical behaviour which involves "treeing" and calling up currents and affects in the surrounding atmosphere. She is the first of her people to be accepted into university in another galaxy, but is forbidden to attend by her family. She sneaks away and heads to university across the galaxy on a living fish-like passenger ship. This book describes a world where technology and nature participate in a shared genesis. The ship sheds her skin after burning in atmosphere during re-entry to planets. Binti calls the third-generation Fish-ship "the finest technology, finest *creature*" the earth has ever produced. This remixing of nature and technics changes the conditions of life and death. In the third book of the trilogy, *Binti: The Night Masquerade,* Binti is absorbed by New Fish (Third Fish's daughter) after being killed in the crossfire of a war. Binti's body is placed in the still growing New Fish, and the microbes that live in the breathing chamber mutate both New Fish and Binti's genes, repairing Binti (and transforming New Fish) into altered beings: part human, part microbe, part New Fish.

Each of these novels raises different insights into species-centric images of life and survival, queering the white and heteronormative images of coupling, reproduction, and sexuality tacitly endorsed in our Western naturecultures. These texts help us interrogate our sense of belonging and the potential of trans-individual sympathetic relationships with what Stacy Alaimo (2016) calls a "wider, more-than-human kin ship network" (p. 30). These texts also highlight new mixtures of technics and nature. Hayles (2012) uses the term "technogenesis" to support the notion that humans and technics have co-evolved together. Technics, in this account, can be anything from stone tools to modern computer processing, but point to an inherent technicity in any worlding process; a technics that is often indifferent to human endeavour. In other words, technics are diversely at work in more-than-human processes. This refusal to start with the assumption that technology is a tool or prosthetic serving human endeavour is essential as we take up the weird and wild worlds of SF science.

Moreover, this perspective helps us bracket the two capacities that are always posited as exceptionally human – technology and language. Simondon (2017) reminds us that technology is poorly understood in social theories that remain focused on the alienation of humans through automaticity and fail to adequately place the human within a techno-ecology. All too often, SF technology is nothing more than an exaggerated version of human thought and human technology, where faster computers and faster rockets advance the human world into a transhuman fantasy. SF narratives that embed technicity into the very fabric of life (alien or otherwise) allow us to think more broadly about ecology.

Annihilation by Jeff VanderMeer (2014) is another powerful example of such SF, as it offers readers a way of imagining *sympathetic* relations with the more-than-human, and shows how such sympathy must engage with affect, thought, and technics, while "holding onto" a non-relational outside. In other words, *Annihilation* describes a sympathetic scientific inquiry that avoids the dangerous celebration of a bland connectivity across difference – it resists an empty endorsement of relationality (Colebrook, 2019a,

2019b) – and instead pursues the complex frictions of violent naturecultures. We contend that the book's focus on "weird" sympathetic relationality helps us explore the ways that inquiry must become "situated" in radical new ways (the term "situated" doesn't quite capture the complexities of imbrication). As Deleuze suggests, sympathy is a bodily struggle that involves affect and the power to be affected (Deleuze & Parnet, 2007, p. 53). This bodily struggle plays out in complex encounters in which various kinds of refusal are negotiated. Such a process involves a metamorphic "making with" that Haraway (2016) calls sympoiesis.

Sympathy: the limits of scientific classification

The word sympathy comes from ancient Greek (sumpátheia) and refers to the state of *feeling together*, derived from a composite of *fellow* and *feeling* (Schliesser, 2015; Spuybroek, 2016). Over the centuries, the notion of sympathy has been used to describe all sorts of activity – everything from contagious yawn catching to cosmological harmony to empathic and even telepathic association (de Freitas, 2018a). Today, in the midst of swarming affect and pan-human stupidity, we advocate for a sympathy that is not a 19th century liberal humanist celebration of the malleability of the white subject (Barnes, 1997), but a more-than-human sympathy that respects and indeed relies upon radical difference and risky diplomacy.

Annihilation involves four female scientists (biologist, anthropologist, psychologist, surveyor) who enter the shimmering Area X to investigate strange mutations and life forms that break with all scientific knowledge. Their mission is to investigate a "localized environmental catastrophe stemming from experimental military research" (p. 94), which we eventually learn has nothing to do with military research. The scientists begin to rapidly *join* or merge with the environment, branching out into territories beyond their own self-maintenance. This eco-SF novel explores the myriad ways the earth is writing itself. The book is separated into five sections (Initiation, Integration, Immolation, Immersion, Dissolution) and is the first book in a trilogy.

The narrator of *Annihilation* is known only as "the biologist." We never learn the names of the other characters. They are simply described as "The Anthropologist" or "The Surveyor" or "The Psychologist," emphasizing their professional identity and their purpose in the team. This nameless habit extends to the biologist's memories about "my husband," underscoring the gender politics of function and relationality. Even the site of the expedition "Area X" goes without a name. And the training that the scientists receive, prior to entry, is also strangely lacking in the kind of detail that might be a mark of the real – they are told that there are four kinds of venomous snakes in Area X but possess no other facts about these. They are told that they are the 12th expedition into the region, and yet the accumulated information regarding the area, including the map, seems limited and superficial. There is a coastline, a lighthouse, various brooks and streams, and forests. They are told that it is a "transitional" ecology and that strange new genetic mutations are appearing, whereby species are adapting in unexpected and accelerated ways. It seems plausible that the old names are simply no longer viable in this transitional ecology.

The biologist joined the team because she was an expert in "transitional environments," but her training with tidal pools and similar bridging ecologies, where cross-pollination and cross-transfers of other material occur, has not prepared her for the eerie and weird transitional species she finds in Area X, where walking for only six miles would take you through forest, swamp, jungle, marsh, beach, etc. She is constantly taking samples from the strange organisms she encounters, extracting tissue to later examine the cellular activity under the microscope at base camp. She later finds brain cells in the walls of a ruined and inverted tower – remarkably human-like cells with non-human irregularities capable of flowering and spuming spores. Her first assumption is that her inhalation of such spores has contaminated the sample (rather than admit that the fruiting walls possess brain tissue). Other data force her to admit a strange hybrid biological process at work across the landscape, where lichen and moss rise up misshapen, forming inexact human limbs and heads and torsos, as though "the natural world around me had become a kind of camouflage" (p. 98) and that "Human lives had poured into this place over time, volunteered to become party to exile and worse" (p. 119). Importantly, the sympathetic transformation is not a one-way exchange, nor does it allow humans to survive in the ways they have always imagined.

Here, sympathy refers to the kind of transindividual experience that allows one to be "moved" by the other but also involves a learning process to actively move with the other. The biologist in Area X exhibits this kind of sympathy as a way of exploring the trans-intimacies that link "species" with/in environments, across previously established differences. Notably, her encounters within this unfamiliar ecology do not aim to cultivate a capacity for "familial feeling" applied further afield in the strangely alien Area X, for that would duplicate the oppression of state-sanctioned family relations onto a queer world (Barnes, 1997). This capacity to be queered is crucial in her burgeoning research practice in Area X, and grounded in her past experiences. She recounts her previous interest in composting life forms, not as a cross-species exchange, companion or otherwise, but as a kind of metamorphic agency without containment, as though her unsentimental disposition towards ecology was also what allowed her to join sympathetically with the alien environment of wildlife in Area X. When the mission in Area X is faltering, and some of the other team members are dead, she is still out seeking to learn with/about in ways that put her own organic life at risk. And yet her training in a still anthropocentric science is what also makes her ill-prepared for "what appeared to be the uncanny" (p. 69). Her encounters with the alien and weird fuel a persistent paranoia, as she imagines herself watched and observed by the strange new wildlife:

> At one point, [a pair of otters] glanced up and I had a strange sensation that they could see me watching them. It was a feeling I often had when out in the wilderness: that things were not quite what they seemed, and I had to fight against the sensation because it could overwhelm my scientific objectivity.
>
> *(VanderMeer, 2014, p. 30)*

We are reminded of the ground-breaking work of Science Studies scholar Evelyn Fox-Keller (1985) and her analysis of the paranoid disposition of certain scientists. Is our biologist encountering an alien existence, or are these animals merely the projections of human desire? Is the environment haunted by humans from previous expeditions? After an encounter with a beast who also seems in the midst of some inner human-like mental contortion, she says, "For several hours, my thoughts turned inward toward explanations for what I had seen: parasites and other hitchhikers of a neurological nature. I was searching for entirely rational biological theories" (p. 17). Here we begin to suspect that past expeditions continue to live in the milieu in other organic forms, having broken out from beyond the sustenance of their own individual survival.

Naturing nature: metamorphosis

But to what extent does this work of fiction help us imagine a scientific empirical practice in "our" world? Latour (2017) demonstrates how science is a highly terrestrial human practice dedicated to encountering various agencies that proliferate and populate the world. In other words, science, at its best, is deeply committed to a mutational "metamorphic zone" and to increasing the number of alien and/or non-human voices that can speak, on their own behalf, as part of a growing political ecology. From this perspective, science opens up environments so that "we" can better sympathize across species and across materiality. For instance, science shows how human bodies are bacterial colonies or that atoms are indeterminate coalitions of micro-particles or that forests mobilize vast signalling networks. These are all examples of how science, at its best, proliferates non-human agencies and remixes the metamorphic zone of life. Notably, Latour's image of science differs from many science denialists who critique science for not being absolute. In other words, denialists cannot trust 20th century science because it is too fragile and tentative in its claims, always hedging, prevaricating, and affirming an onto-epistemology of indeterminacy, contingency, and uncertainty. While new materialisms celebrate these qualities of contemporary science, those who are distrustful often critique what they now see as a postmodern science that is too biased and relativistic! (Latour, 2018).

In contrast, Latour argues that scientific practice actually expands a more inclusive *realism* through acknowledging (often making visible to humans) the myriad entangled agencies and more-than-human power networks that compose "metamorphic matter." Latour presents a science that is meant to create opportunities for new forms of non-human sign-making, thereby multiplying political and material agencies across an ecology. This is an image of science that increases the ontological pluralism, expanding the alliances, and ultimately assembling another more-than-human political body – a "science that populates the world with diverse non-human agencies (gluons, amoebas, gravitational waves, etc.), according to an immanent ontology *that is no less realist for being thoroughly situated*." (de Freitas, 2020). It may well be that the biologist's methods of inquiry in *Annihilation* are not adequate to the new ecology she encounters, but the point of science is to pursue that encounter nonetheless, uncovering previously unknown agencies and composing a risky diplomacy (Latour, 2017).

Apparatus play a pivotal role in Latour's approach to scientific method and have been theorized in much qualitative research that draws on new materialisms, demonstrating the significance of an apparatus in the outcome of a scientific experiment (de Freitas, 2017b; Truman, 2016a). Notably, the team of female scientists in *Annihilation* brings no current technology, nothing digital or satellite-based, no video or complex measuring instruments into their investigations of Area X. There is some sense that such technology might be too easily absorbed and appropriated by the unknown forces at work. Rather, they use old-school microscopes to examine samples. They are given paper notebooks and told they must document in detail all their observations and findings. Significantly, these field notes eventually figure very prominently in the narrative, when the biologist finally enters the Lighthouse to find a two-story heap of stacked notebooks from previous scientific missions. These accounts by previous scientists form a midden of written refuse, a testimony to past inquiry, abandoned and forgotten – indeed, she finds more field notes than could have possibly been left by only 12 expeditions. She picks through the pile, reading here and there, and learning about the various teams who have come before.

What do these notebooks testify to? Have they increased our understanding of Area X? Are they accounts of a learning process or filled with a violent refusal to merge and engage in trans-intimacies? The midden of scientific notebooks underscores the limitations of observational science and conventional forms of documentation – a mulched, sloppy pile of unreadability. If this pile is the accumulated knowledge of Area X, it seems to add up to a refuse heap. Or perhaps more accurately, it emphasizes how scientific observation, as a canonical empirical activity, may in fact be a very personalized accounting of oneself, a form of reflexive *composting*. The novel thus helps us imagine a new form of observation that might be theorized around this insight – a form of observation that is essentially a generative process of *composting*. This moves us away from social constructivist readings of observation that undervalued the force of matter, to a theory of observation that reckons with co-relational and de/creative processes.

The materiality of language

Poststructuralist literary scholars have for decades discussed everything from the author-reader co-creation of meaning, to subject dissolution, to infinite deferral of meaning (e.g. Barthes, Foucault, Derrida); and more recently feminist new materialists have drawn attention to the flesh and materiality of text and how such narration might materially transform readers (Grosz, 2004; Truman, 2016a, 2016b, 2019b). In a related fashion, *Annihilation* draws attention to the materiality of writing and textuality.

Notably, a linguist was trained for the expedition, but suddenly decided to drop out. Why her sudden departure? This is the first clue that the materiality of language plays a key role in this novel, and that the human science of linguistics is ill-prepared to reckon with what the team will find in Area X. Indeed, the novel contests our limited grasp of the technicity of language, overhauling the very notion of communication and signal; the novel complicates our investments in separating meaning from matter, and text from context.

The four scientists encounter a living language inside Area X, while investigating a strange architectural "feature" of the otherwise "natural" environment, a remnant of a past "built environment." They uncover an open hole and descend into what seems to be an inverted tower. The tower is a mixture of human invention (cement) and the carapaces of organic life (seashells), immediately signalling the fact that human technics are always already mobilizing (and being mobilized by) the agencies (and deaths) of non-human life. The tower, made of stone and coquina, is a key image for the story, an image that stands for the dark matter within (or beyond) the erect lighthouse beacon to which it is thematically and plot-linked. The biologist takes a huge interest in the inverted tower and is drawn below into its underground twisting path.

Once inside the tower, the scientists encounter what the biologist will name the *Crawler*. Notably, this is the first use of a proper name in the novel, in that she uses the term to name what seems to be a creature. As they descend the stairwell into the tower, they first spy, at about shoulder height, a glistening green vine clinging to the inner wall and heading downward into the spiral staircase. As the biologist examines it more carefully, the vines resolve into tiny words, a cursive lettering that rises off the wall. She is surprised to understand the words, and recognize them as human words, filling her with an intertwined sense of elation and dread. In the vine, she reads: "*Where lies the strangling fruit that came from the hand of the sinner I shall bring forth the seeds of the dead to share with the worms that …*" (italics in original, p. 23). The anthropologist asks of the words, "what are they made of?" which draws our attention to their strange materiality while we simultaneously absorb the sound, syntax, and semantics of the ominous unfolding refrain. The words "are made of" green fernlike moss or fungi or some other kind of eukaryotic organism, writhing with curling filaments: "A loamy smell came from the words along with an underlying hint of rotting honey" (p. 24). Here the earth seems to be writing itself.

The words themselves are *fruiting bodies* as well as *semantic-syntactic forms*. The writing on the wall is both *matter* and *meaning*, and the biologist feels the phrases infiltrating her mind in unexpected ways, "finding fertile ground" to grow and mutate. She leans in to get a closer look and accidentally inhales a tiny spray of golden spores that spews out from the vine. The scientists pursue the cursive curly filaments, quivering alive and foreboding. They find tiny "hand-shaped parasites" that live among the words, active creatures that sustain the words, while destroying them as well. The novel directs out attention to the *composting of words, as though they were both garbage and nourishment*:

> My samples told a series of cryptic jokes with punch lines I didn't understand. The cells of the biomass that made up the words on the wall had an unusual structure, but they still fell within an acceptable range. Or, those cells were doing a magnificent job of mimicking certain species of saprotrophic organisms. I made a mental note to take a sample of the wall from behind the words. I had no idea how deeply the filaments had taken root, or if there were nodes beneath and those filaments were only sentinels. The tissue sample from the hand-shaped creature resisted any interpretation, and that was strange but told me nothing. By which I mean I found no cells in the sample, just a solid amber surface with air bubbles in it. At the time, I interpreted this as a contaminated sample or evidence that this organism decomposed quickly. Another thought came to me too late to test: that, having absorbed the organism's spores, I was causing a reaction in the sample.
>
> *(p. 71)*

The meaning and materiality of the words absorbed by the biologist begin to affect her both chemically and psychologically. Like the other SF stories mentioned earlier, the transaction is not one directional, and transgresses conventional borders between meaning and matter.

Unreliable scientific narrators

The literary device "unreliable narrator" is used effectively in this novel to keep readers suspicious about the veracity of any given account. Authors use unreliable narrators to create stories that are radically open to unravelling, breaking up the sometimes oppressive controlling nature of the story to contain events. The device goes some way to countering the kind of totalizing tendency in many current SF films where humanity and world are rescued (usually by a white man) for a proper humanity and "people yet to come" (Colebrook, 2019a, 2019b). The unreliable narrator may be wilful in their deceit or simply ignorant of the larger forces that are at work in the story world. Examples of unreliable narrators are everywhere in literature. Authors use cues to help readers know that the perspective is entirely situated and perhaps flawed. The biologist of *Annihilation* is an unreliable narrator, which keeps the reader wondering whether this is an accurate and true account of an actual expedition into a strange new ecology, or whether this is the dream or delusion of someone who is under hypnosis. This unreliability is a crucial element in our reading of the book and is pivotal for how it sets up new STEM imaginaries in post-truth conditions. In *Annihilation*, we are cued early on to the narrator's limited understanding of the events she recounts. The first pages have us wondering if the entire account is a hypnotic simulation, entirely built up from the psychological concerns of the main character, undergoing an experiment of some kind, especially as the term "transference" is used to describe the entry into the Zone (p. 5).

The use of an unreliable *scientific* narrator is a significant device for problematizing the biological knowledge that our main character brings to the new alien ecology. "...all of this speculation is incomplete, inexact, inaccurate, useless...our instruments are useless, our methodology broken, our motivations selfish" (VanderMeer, 2014, pp. 192–193). The fact that the biologist is unreliable resonates with insights from Science Studies that emphasize the partial, situated, and contested nature of scientific knowledge (Subramaniam, 2014). These insights are not meant to undermine science but point towards its complexity and ongoing evolution as a practice. As narrator, the biologist speaks in first person, delivered in an almost affectless neutral voice, posing as the detached objective scientist, a position that she cannot possibly maintain in this strange environment. The almost theatrical performance of scientific neutrality and rationality is under siege in the posthuman ecology, where new mixtures of mind and matter emerge. There are no isolated and transcendent points or positions of objective observation in Area X. Indeed, the biologist's scientific methods are increasingly transindividual, as she merges with the environment. This kind of immersive transformation – the very opposite of classical objectivity – is precisely what is entailed in the kind of sympathetic relationality discussed above, in which a shared bodily struggle sustains an assembling of divergent agencies. We propose that her research methods become a kind of posthuman ethnography as she explores this radical mutational ecology where ethno and eco mix.

We want to think further with unreliability and propose that an unreliable first-person scientific narrative voice is crucial in helping the reader begin to imagine a different socio-material research practice. At the same time, we are fully conscious of how this unreliability is fuelling anti-science sentiment on the far-right. Rather than read her affectless tone as resistance to change, we defend her impersonal "naturalist" voice as an indicator of her eager and diplomatic trans-species relationality. The biologist's disposition is what allows her to engage with the posthuman ecology, rather than with some emotional and sentimental image of the holobiont (Schuller, 2018). This kind of determination puts both self and world at risk; scientific experiments with consequential meaning often involve this kind of indifference to survival, when the scientist is said to pursue truth without concern for their own safety (Johnson, 2008). Although in some cases this is the heroic packaging of the (male, able, hetero, white) scientist, we believe in this case it's worth considering the risks involved in empirical research, as a form of truth-seeking under post-truth conditions. In *Annihilation*, for instance, the psychologist is an untrustworthy scientist, as she leads the team through attempts at hypnosis. If the scientists are simply subject to some sort of hypnosis, why should we trust their knowledge-seeking methods? How can they achieve any sort of agential realism if they are being manipulated by the untrustworthy psychologist? But the biologist, despite being unreliable, is trustworthy. In this case, the biologist develops a resistance to the hypnotic refrain, precisely because her reckless empirical methods brought her into intimate contact with the spores of the Crawler. Attending to these complex tensions between being *trustworthy* but *unreliable* might help us grapple with science denialism and new empiricisms under post-truth conditions.

Kortekallio (2019) suggests that the biologist's exposure to the Crawler's infectious spore-writing changes everything: Citing William S. Burroughs, she states, "After this … I realize that everything surrounding me is undoubtedly and uncontrollably alive, and as a consequence lose the sense of being the center of my world" (p. 61). This de-centring of the human and their assumed control creates a feeling of "becoming instrument" that both the reader and the biologist experience, as they feel a resonance, impressibility, and the disturbing but also enhancing feeling of "being played" by both the unreliable narrator and the highly unpredictable environment (Kortekallio, 2019). Notably, this is not about identification and a process whereby the reader is "becoming" the first-person narrator – rather Kortekallio argues that we maintain an awareness of the fictional nature of the text while aligning with the first-person narration, much like other posthuman narration (Clarke, 2008). She equates this strategy with the feminist new materialist scholars' notion of "thinking-with" (i.e. Haraway, 2008).

In becoming-instrument and thinking-with Area X, the reader begins to appreciate that the "viscosity" of the ecology is so totalizing that one cannot distance oneself from it. In other words, the reader tags along with the scientists who *undergo* inquiry, as much as they *perform* inquiry. This kind of ecological reading is not only attunement, but more like being played (with all the betrayal that such a term implies) by the environment. Granted, there are multiple ways of attuning, and Kortekallio (2019) explains how "being played" also entails the anxiety of influence that might lead to defensiveness, and perhaps also to the fetishizing of familiar established forms of inquiry.

Resisting a sentimental relationality

As a rumination on forms of inquiry in environmental disaster zones, the novel *Annihilation* has direct relevance to our own climate crisis. The need to grapple with the reality of these "end times" has been taken up in the humanities and is directly linked to the future of science (Donowski & Viveiros de Castro, 2017). To that end, we see at least two kinds of end-times and annihilation in Area X. First, there is the annihilation that the psychologist, as team leader, is meant to offer the team if they need it – she uses the word *annihilation* as one of the key hypnotic words meant to induce immediate suicide when uttered. This is an all-too human annihilation, a suicide that serves the controlling forces outside of Area X. There is a second kind of annihilation in the novel that involves a slow merging with the genetic tissue of other organisms, that is, the surrender to an environmental force within the relational milieu that is larger than the individual organism's will to survive. This is a kind of metamorphosis and dissolution. This kind of annihilation transcends and destroys speciation. The biologist is herself undone by the transitional environment, but her annihilation is a slow dissolution, through contagion or infection, when the spores she inhales infiltrate the body and transform her into a flame of new life.

Creative mimesis plays the ultimate de/creative role in Area X, where annihilation is achieved through the *environmental technics of mimicry* and the proliferation and permutation of minute characteristics. The Crawler operates through a kind of mimicry using the biologist's and others' thoughts. This relentless assimilating mimicry is like a "thorn" in the side of the world, mutating at scales and speeds beyond human perception. The biologist muses:

> It creates out of our ecosystem a new world, whose processes and aims are utterly alien – one that works through supreme acts of mirroring, and by remaining hidden in so many other ways, all without surrendering the foundations of its *otherness* as it becomes what it encounters.
>
> *(VanderMeer, 2014, p. 191)*

This book thus offers an image of non-narcissistic mixtures of matter-meaning and posthuman "naturalist" forms of inquiry. The novel helps us imagine a world where sociality and materiality are de/creatively imbricated beyond human interest; the Crawler's spore-script is not intended for future human expeditions – the words are not intended as messages to warn the incoming. This is not a sentimental connectivity in Area X. Communication is an environmental more-than-human activity, rather than a willed message from source to human receiver. The biologist must learn how to recognize the experimental indifference of the world to human endeavour. In the biologist, we glimpse the burgeoning of a posthuman science, invested in a new kind of impressibility, but still clinging to her conventional scientific training as she brings forth new methods of encountering an outside: "My sole gift or talent, I believe now, was that places could impress themselves upon me, and I could become a part of them with ease" (p. 110).

Importantly, the book resists a sentimentality about such radical relationality and impressibility. In her brilliant analysis of race, sex, and science in the 19th century, Kyla Schuller (2018) suggests that past investments in a sentimental biopower of "impressibility" have often been a means for separating the "sensitive" civilized white and heteronormative subject from the abject other. We are cautious and wary that emphasis on sympathy and affect and plasticity inherits "the political legacies of liberal humanism … [and may] unwittingly recapitulate the conceptual apparatus of the biopolitics of feeling" (Schuller, 2018, p. 11). For this reason, it's essential to recognize the "hierarchical intimacies" that structure the relational space of Anthropocene life, and to seek "a sympathetic coordination [that] is not a bland alignment between different agents, nor an identification amongst group members, nor the creation of a unified homogeneous assemblage, but is a coordinated assembling of heterogeneous agencies." (de Freitas, 2018a, p. 89). Such rethinking of agency, and its correlate concept of causality, is no easy matter. Latour argues that conventional renderings of science rest on a faulty concept of causality, which ascribes all the agency to the cause and none to the effect:

> Shedding our attachment to causes that literally control all the action would entail queering time as we know it. This demands a new kind of causality which tracks the

reciprocal mutations of the metamorphic zone, linking animacies and agencies across times in new ways.

(de Freitas, 2019, p. 6)

Afterthoughts on the future of empiricism

The plot and poetics of SF novels like *Annihilation* immerse us as readers into worlds where more-than-human sympathy and posthuman ecologies are *part* of the fabric of reality. Such SF literary realities invite us to imagine both the scientific method and the socio-biological body otherwise. In search of new forms of Anthropocene inquiry, they help us make sense of new corporeal metamorphosis and mind-body mixtures. These texts help researchers bracket the "anthropos" so as to understand the multiple ways in which diverse "peoples" inherit the climate crisis. We emphasize that our proposal does not defend the same old euro-centric sciences, nor does it necessarily imply human survival in the Anthropocene – again, we are not advocating for a sentimentalism or biopolitics of feeling that imprints whiteness or humanity onto everything that was once alien or other. Stories like *Annihilation* underscore both the plurality of earthly end-times as well as grand cosmic scales of destruction.

We close by asking: To what degree are these SF texts relevant to current attempts to remix the social and physical sciences in concrete empirical practice? To that end, the historian of Science Myra Hird and critical geographer Kathryn Yusoff offer insight into burgeoning new research methods in their study of the strange new naturecultures of a toxic superfund site where microbial alliances are formed with minerals, metals and human health experts, to inform an "emerging science of mineralogical evolution" (Hird & Yusoff, 2019, p. 265). The Butte superfund site was an abandoned open pit copper mine, later flooded in order to "contain" the disaster, thereby forming a metal-laden lake. The lake attracted migrating birds and became a local compost site of mass death and bacterial innovation after flocks of Canadian Geese mistook the shiny surface for a pristine lake. Hird and Yusoff describe the inhuman trajectories that animate the toxic metal lake, as well as the humans who come in search of new bacterial actants that might cure human cancer. We suggest here that the authors create a *posthuman ethnography*, or a geostory, making visible to the reader the complex agentic and metamorphic socio-material zone of a posthuman ecology. This is not a work of science fiction, but the practices described therein share many qualities with those we have found in SF. The authors present an empirical inquiry that matches the Latourian description of science we cited earlier, where the drama unfolds through a set of frictional encounters between human and non-human characters, a drama that is *ground in the relational performances of those characters*.

We believe this work on the emergent mineral-microbrial alliances and "inhuman trajectories" (between life and nonlife) at this ecological disaster site offers a glimpse into the future of Anthropocene inquiry. The example of Hird and Yusoff gestures towards a form of inquiry into naturecultures which are somewhat indifferent to the survival of the bounded human organism, and yet this is also a site where new sympathetic and symbiotic relationships flourish. Their method tracks a more-than-human belonging and becoming that entails more than simply autopoetic self-maintenance, differing from the individualistic-organism image we've inherited from systems theory (Clarke, 2017).

In this chapter, we have used literary SF as a way of exploring scientific imaginaries in the Anthropocene, and hope to have demonstrated how SF may be helpful to researchers who are seeking ways to complicate empirical methods and methodology. SF images of new empiricism can spur us on to act with more-than-human sympathy in posthuman ecologies that demand a pluralist and plastic sense of worldly belonging. Books like *Annihilation* offer a glimpse into otherworldly earthly terrains, much like the ecologies in which we actually find ourselves today. We've shown how the novel's plot, imagery, and character development help us explore questions about the future of posthuman empiricism, in a world where new matter-mind mixtures and in/organic variation proliferate.

Notes

1 The Anthropocene can be described as the epoch in chronological time when human impact on earth has reached a geologic scale – and the implications of this on the future of life on earth.
2 Even as we write this, the term "ecology" has become almost emptied out in its over-application (Neyrat, 2017). Latour (2017, 2018) notes that ecology has, in fact, failed us in fundamental ways, as a field dedicated to more-than-human relationality. And yet, at this point, it still indexes a set of powerful ideas (Hörl, 2018).

References

Alaimo, S. (2016) *Exposed: Environmental politics and pleasures in posthuman times*. Minneapolis: University of Minnesota Press.
Alumkal, A. (2017). *Paranoid science: The Christian right's war on reality*. New York: New York University Press.
Bahng, A. (2017). Plasmoidial improprieties: Octavia E. Butler, Slime molds, and imagining a Femi-Queer commons. In C. Cipolla, K. Gupta, D. A, Rubin, & A. Willey (Eds.), *Queer feminist science studies* (pp. 310–326). Seattle: University of Washington Press.
Barad, K. (2007). *Meeting the universe half way: Quantum physics and the entanglement of matter and meaning*. Durham, NC: Duke University Press.
Barnes, E. (1997). *States of sympathy: Seduction and democracy in the American novel*. New York: Columbia University Press.
Braidotti, R. (2013). *The posthuman*. Cambridge, MA: Polity.
Braidotti, R., & Hlavajova, M. (2018). *Posthuman glossary*. London: Bloomsbury Press.
Butler, O. (2000). *Lilith's brood*. New York: Grand Central Publishing.
Chakrabarty, D. (2014). Climate and capitalism: On conjoined histories. *Critical Inquiry*, 42, 1–23.
Clarke, B. (2008). *Posthuman metamorphosis: Narrative and systems*. New York: Fordham University Press.
Clarke, B. (2017). Planetary immunity: Biopolitics, Gaia theory, the holobiont, and the systems counterculture. In E. Hörl (Ed.) (2018), *General ecology: The new ecological paradigm* (pp. 193–216). London: Bloomsbury.
Clough, P. T. (2009). The new empiricism: Affect and sociological method. *European Journal of Social Theory*, 12, 43–61.
Colebrook, C. (2014) *Death of the posthuman: Essays on extinction* (vol. 1). Ann Arbor, MI: Open Humanities Press.
Colebrook, C. (2019a). A cut in relationality: Art at the end of the world. *Angelaki: Journal of the Theoretical Humanities*, 24(3), 175-195.
Colebrook, C. (2019b). *Unthinkable extinction: Cinematic time and the panorama of history. Alienocene: Journal of the First Outernational*. Available at https://alienocene.com/2019/06/19/unthinkable-extinction/

Coole, D., & Frost, S. (2010). *New materialisms: Ontology, agency, and politics*. Durham, NC: Duke University Press.
Danowski, D., & Viveiros de Castro, E. (2017). *The ends of the world*. (Trans. R. Nunes). Cambridge: Polity Press.
de Freitas, E. (2017a). Non-human findings from the laboratory of speculative sociology. *Minnesota Review*, 88, 116–126.
de Freitas, E. (2017b). Karen Barad's quantum ontology and posthuman ethics: Rethinking the concept of relationality. *Qualitative Inquiry*, 23(9), 741–748.
de Freitas, E. (2018a). Love of learning: Amorous and fatal. In S. Bignalle & R. Braidotti (Eds.), *Posthuman ecologies: Complexity and process after Deleuze*. (pp. 87-104). Lanham: Rowman & Littlefield.
de Freitas, E. (2018b). The biosocial subject: Sensor technology and worldly sensibility. *Discourse: Studies in the Cultural Politics of Education*, 39(2), 292–308.
de Freitas, E. (2019). Science studies and the metamorphic multiple earth: Bruno Latour's risky diplomacy. *Cultural Studies – Critical Methodologies*, 20, 203–212. Special Issue on Rethinking social inquiry in the wake of Science Studies: Transdisciplinary pursuits in times of climate change, information flows, and fading democracies (Ed. E. de Freitas & J. Weaver).
de Freitas, E. (2020). Why trust science in trickster world of absolute contingency? *Cristal: Critical Studies in Teaching and Learning*, 8. https://doi.org/10.14426/cristal.v8iSI.278
de Freitas, E., & Curinga, M. X. (2015). New materialist approaches to the study of language and identity: Assembling the posthuman subject. *Curriculum Inquiry*, 45, 249–265.
de Freitas, E., Lupinacci, J., & Pais, A. (Eds.). (2017). Science and technology studies x educational studies: Critical and creative perspectives on the future of STEM education. *Educational Studies: A Journal of the American Educational Studies Association*, 53(6), 551–559.
de Freitas, E., & Truman, S. (2019). Science dis/trust in the end-times: Bruno Latour and *The Three-body Problem. Paper Presented at the Association for Philosophy and Literature Annual Conference*, Klagenfurt, Austria.
de Freitas, E., & Weaver, J. (Eds.). (2019). Rethinking social inquiry in the wake of Science Studies: Transdisciplinary pursuits in times of climate change, information flows, and fading democracies. *Cultural Studies – Critical Methodologies*, 20, 195–202.
Delanda, M. (2015). *Philosophical chemistry*. London: Bloomsbury Academic.
Deleuze, G., & Parnet, C. (2007). *Dialogues II: Revised edition*. New York: Columbia University Press.
Drummond, C., & Fischhoff, B. (2017). Individuals with greater science literacy and education have more polarized beliefs on controversial science topics. *Proceedings of the National Academy of Sciences*, 114(36), 9587–9592. https://doi.org/10.1073/pnas.1704882114
Fox-Keller, E. (1985) *Reflections on gender and science*. New Haven, CT: Yale University Press.
Grosz, E. (2004). *Nick of time*. Durham, NC: Duke University Press.
Haraway, D. (2008). Companion species, mis-recognition, and queer worlding. In N. Giffney & M. J. Hird (Eds.), *Queering the non/human* (pp. xxiii–xxvi). Burlington: Ashgate.
Haraway, D. (2013). SF: Science fiction, speculative fabulation, string figures, so far. *Ada: A Journal of Gender, New Media, and Technology*, 3.
Haraway, D. (2016). *Staying with the trouble: Making kin in the Chthulucene*. Durham, NC: Duke University Press.
Hayles, K. (2012). *How we think: Digital media and contemporary technogenesis*. Chicago, IL: The University of Chicago Press.
Hird, M., & Yusoff, K. (2019). Lines of shite: Microbial-mineral chatter in the Anthropocene. In R. Braidotti & S. Bignalle (Eds.), *Posthuman ecologies: Complexity and process after Deleuze* (pp. 265–82). New York: Rowman & Littlefield.
Hörl, E. (Ed.). (2018). *General ecology: The new ecological paradigm* (pp. 172–174). London: Bloomsbury.
Johnson, G. (2008). *The 10 most beautiful experiments*. New York: Alfred Knopf.

Keeling, K. (2019). *Queer times, black futures*. New York: New York University Press.
Kortekallio, K. (2019). Becoming-instrument: Thinking with Jeff VanderMeer's *Annihilation* and Timothy Morton's *Hyperobjects*. In S. Karkulehto, A. Koistinen, & E. Varis (Eds.), *Reconfiguring human, nonhuman and posthuman in literature and culture* (pp. 57–75). London and New York: Routledge.
Latour, B. (2017). *Facing Gaia: Eight lectures on the new climatic regime*. Cambridge: Polity Press.
Latour, B. (2018). *Down to earth: Politics in the new climatic regime*. Cambridge: Polity Press.
Lewandowsky, S., Gignac, G. E., & Oberauer, K., (2013). The role of conspiracist ideation and worldviews in predicting rejection of science. *PLoS ONE*, 8 (10), e75637.
MacLure, M. (2013). Researching without representation? Language and materiality in post-qualitative methodology. *International Journal of Qualitative Studies in Education*, 26(6), 658–667.
Moore, J. W. (2016). *Anthropocene or capitalocene? Nature, history, and the crisis of capitalism*. Oakland, CA: Kairos Books.
Morton, T. (2016). *Dark ecology: For a logic of future coexistence*. New York: Columbia University Press.
Neyrat, F. (2017). Elements for an ecology of separation: Beyond ecological constructivism. In E. Hörl (Ed.), *General ecology: The new ecological paradigm* (pp. 193–216). London: Bloomsbury.
Parisi, L. (2017). Computational logic and ecological rationality. In E. Hörl (Ed.) (2018), *General ecology: The new ecological paradigm* (pp. 75–100). London: Bloomsbury.
Povinelli, E. A. (2018). Mother earth: Public sphere, biosphere, colonial sphere. *Eflux*, June #92. Available at https://www.e-flux.com/journal/92/204673/mother-earth-public-sphere-biosphere-colonial-sphere/
Protevi, J. (2013). *Life, war, earth: Deleuze and the sciences*. Minneapolis: University of Minnesota Press.
Ruppert, E., Law, J., & Savage, M. (2013). Reassembling social science methods: The challenge of digital devices. *Theory, Culture, Society*, 30(4), 22–46.
Schliesser, E. (Ed.). (2015). *On sympathy*. Oxford: Oxford University Press.
Schuller, K. (2018). *The biopolitics of feeling: Race, sex and science in the 19th century*. Durham, NC: Duke University Press.
Simondon, G. (2017). *On the mode of existence of technical objects*. (Trans. C. Malaspina & J. Rogove). Minneapolis, MN: Univocal.
Snaza, N., Sonu, D., Truman, S. E., & Zaliwska, Z. (2016). Introduction: Re-attuning to the materiality of education. In N. Snaza, D. Sonu, S. E. Truman, & Z. Zaliwska (Eds.), *Pedagogical matters: New materialisms and curriculum studies*. (pp. xv–xxiv). New York: Peter Lang.
Springgay, S., & Truman, S. E. (2018). On the need for methods beyond proceduralism: Speculative middles, (in)tensions, and response-ability in research. *Qualitative Inquiry*, 24(3), 203–214.
Spuybroek, L. (2016). *The sympathy of things: Ruskin and the ecology of design*, 2nd edition. New York: Bloomsbury Academic Press.
Stengers, I. (2018). *Another science is possible*. Trans. Stephen Muecke. Polity Press.
St. Pierre, E. A. (2016). The empirical and the new empiricisms. *Cultural Studies ↔ Critical Methodologies*, 16(2), 111–124. https://doi.org/10.1177/1532708616636147
Subramaniam, B. (2014). *Ghost stories for Darwin: The science of variation and the politics of diversity*. Urbana-Champaign: University of Illinois Press.
Suvin, D. (1972). On the poetics of the science fiction genre. *College English*, 34(3), 372–382.
Todd, Z. (2016). An Indigenous feminist's take on the ontological turn: 'Ontology' is just another word for colonialism. *Journal of Historical Sociology*, 29(1), 4–22.
Truman, S. E. (2016a). Intratextual entanglements: Emergent pedagogies and the productive potential of texts. In N. Snaza, D. Sonu, S. E. Truman, & Z. Zaliwska (Eds.), *Pedagogical matters: New materialisms and curriculum studies* (pp. 91–108). New York: Peter Lang.
Truman, S. E. (2016b). Becoming more than it never (actually) was: Expressive writing as research-creation. *Journal of Curriculum and Pedagogy*, 13(2), 136–143.
Truman, S. E. (2017). Speculative methodologies & emergent literacies: Walking & writing as research-creation (Unpublished doctoral dissertation). University of Toronto, Canada.

Truman, S. E. (2019a). Feminist new materialisms. In P. Atkinson, S. Delamont, A. Cernat, J. W. Sakshaug, & R. A. Williams (Eds.), *SAGE research methods foundations*. https://doi.org/10.4135/9781526421036808740

Truman, S. E. (2019b). SF! Haraway's situated feminisms and speculative *Fabulations* in English class. *Studies in Philosophy and Education*, Special Issue on Donna Haraway. https://doi.org/10.1007/s11217-018-9632-5

Tsing, A. (2017). *The mushroom at the end of the world: On the possibility of life in capitalist ruins*. Princeton, NC: Princeton University Press.

VanderMeer, J. (2014). *Annihilation*. New York: Farrar, Straus and Giroux.

Wendt, A. (2015). *Quantum mind and social science: Unifying physical and social ontology*. Cambridge: Cambridge University Press.

Willey, A. (2016). *Undoing monogamy: The politics of science and the possibilities of biology*. Durham, NC: Duke University Press.

Wilson, E. (2015). *Gut feminism*. Durham, NC: Duke University.

Yusoff, K. (2019). *A billion black Anthropocenes or none*. Minneapolis: University of Minnesota Press.

11

IS THERE AN OUTSIDE TO RELATIONALITY IN A TRICKSTER WORLD OF ABSOLUTE CONTINGENCY?

Introduction

What is entailed when we pursue metamorphic couplings across conventional divides of matter-meaning, using both empirical and speculative techniques? What kind of theoretical tensions and philosophical dilemmas come with a commitment to new forms of *empiricism and speculation*? In this chapter, I dwell on the problematic but also productive tensions that arise in attempts to remix the physical and social sciences, attending to somewhat uncomfortable affinities with various kinds of philosophical naturalism and realism. I am interested in how the very notion of relationality is both at work and at risk in various contemporary materialisms (agential, speculative, promiscuous, etc.). Claims to relationality often entail an overreach of some kind, as Claire Colebrook (2019a) suggests, calling out "the implicit moralism of this posthuman relationality" which often seems intent on rescuing a humanity through spreading its relations, finding itself again and again in its monstrous inventions and its alienated others (Colebrook, 2019a, p. 175). Celebrations of relationality can be rather self-serving ways of depoliticizing frictional encounters, and failing to accept limits and incommensurables. Refusing to let being "be" without relation is a way of colonizing the world with Humanist desire, part of the Kantian legacy of correlationism, discussed in more detail in Meillassoux (2008). The fruitful nature of a term like relationality, and in particular the way it directs our attention to plasticity and processes of becoming, often belies the ways in which it is lived in radically disproportionate ways.

I take Kyla Schuller's cautions (2018) that the capacity to be affected – that is, the capacity to be 'in relation' and transformed – is often associated with a Eurocentric sentimental biopower of "impressibility" formulated in the 19th century, a form of biopower that effectively functioned to separate the "sensitive" civilized from the abject other. And so I am cautious and wary that emphasis on relational affectivity inherits "the political legacies of liberal humanism … [and may] unwittingly recapitulate the conceptual apparatus of the biopolitics of feeling" (Schuller, 2018, p. 11). I say this as someone who is nonetheless committed to the project of relational ontology. Insofar as new materialisms

DOI: 10.4324/9781003538769-11

seek trans-intimacies that merge "species" with/in environments, the risk of spreading white/fe/male/human "familial feeling" further afield is always there, with the associated risks that such a project involves infecting other life with our violence, and potentially duplicating the oppression of state-sanctioned human relations (Barnes, 1997).

This chapter dwells on the concern that "relationality" might in itself be misconceived or misapplied. I consider proposals for an "outside" to relationality, either through the power of speculation or contingency. I venture into this topic by exploring the distinctive power of mathematical speculation and its role in warranting truth and trust in new forms of ecological science. My writing on mathematics is not included in this collection of essays (see de Freitas, 2023, 2025a), but this chapter offers some crossover, drawing on my keynote at the 10th Annual New Materialism conference, in Cape Town, South Africa, in December 2019. In this chapter, I focus on the novel by Cixin Liu called *The Three-Body Problem*, and offer additional analysis to that found in my jointly published work with Sarah Truman (de Freitas & Truman, 2020a, 2020b). I follow many others in Science Studies who have taken up speculative fiction to think through the many pasts, presents, and futures of scientific method. What follows here is an attempt to push this conversation into some uncomfortable philosophical terrain –I will stay focused on the nature of "the speculative" and the limits of a "speculative" science in relational ontologies, as a way to think the *outside*.

The role of speculative thought is crucial to empiricism. In addition to being the imaginative force of thinking-otherwise (Åsberg et al., 2015), speculative thought becomes also that which breaks up and disunifies science – in other words, speculative thought breaks up the monolithic unity of science, and plugs into an ontological diversity that refuses totalizing capture. In this way, speculation multiplies the possible epistemic positions without "reverting to relativism or constructivism" (James, 2019, p. 149). Significantly, following Kragh (2015), mathematics is often the speculative force pushing science outside itself. This opens up science to promiscuous realisms that entail a limitless multiplicity of the real (Dupré, 1993). According to this approach, one can affirm the epistemological disunity of science *and* ontological disorder of matter. And yet, as I discuss below, how does one affirm that and grapple with contemporary science deniers who are concerned with its apparent relativism and changing consensus? How should we understand the disunity of science, as we seek a planetary perspective under new climatic regimes? What if mathematics is more than a compliant maidservant to the sciences, and instead actually fuels that disorder?

I turn to speculative fiction because it pursues the promiscuous power of the imagination to think otherwise, often breaking with the assumptions of contemporary science (Bahng, 2017). Promiscuity is used here as a positive queering of conventions of practice and a way of contesting ontological assumptions about the nature of matter. Elsewhere, I have written about the use of SF to explore mathematical imaginaries, and have developed the idea of "xenomathematics" to emphasize the alienating nature of mathematical thought (de Freitas, 2024, 2025b). Does the mathematical gesture potentially break with science or help us push past the limits of our current knowledge, toward, dare I say, more-than-human inclusive worlds? Certainly, the fantasy and horror writer H.P. Lovecraft used mathematics, and specifically weird geometries, to characterize strange worlds beyond our own. But why would we pursue a world beyond description and relationality? Or should we resist pursuit and just try and open up to its existence? Lothian (2018) reminds us, looking back at old European speculative fiction, that even justice-oriented authors whose aim was to use SF to "halt and divert" various kinds of injustice, often crafted narratives that nonetheless supported heteronormative and white supremacist imaginaries (p. 56). Rieder (2008) similarly recounts the ways in which SF is invested in the human capacity for colonialism. Speculative thought is not innocent (Colebrook, 2019b). Keeling (2019) explores in particular the difference between *corporate* speculative investments (serving to further extractive industries) and SF literature and film that operate through a more "radical imagination" of falsification and fantasy. She deploys various queer formulae to refuse corporate imagined futures and draws on Gilles Deleuze and Édouard Glissant, among others, to further cultivate a poetics of speculative thought. But how do we know when our speculative gestures are serving corporate interests and when they are serving social justice? My aim here is to open up some of these theoretical tensions, in post-truth conditions, by drawing attention to some of the ways that trust and speculation are entangled under chaotic climatic regimes.

Science dis/trust

Trust is a complex relation involving affective, cognitive, practical, emotional, and absurd facets. Public trust in science in the US has significantly diminished from the post-war days of naïve enthusiasm (Alumkal, 2017). Distrust of climate science, in particular, is correlated with libertarian free market advocates (Gauchat, 2012). But often this research begs the question – what is the image of science that is assumed in such research? In exploring distinctions between science distrust and science denialism, Hansson (2017) defines science as "the practice that provides us with the most reliable (i.e. epistemically most warranted) statements that can be made, at the time being, on subject matter covered by the community of knowledge disciplines." (p. 63). This definition emphasizes truth warranting and epistemic authority, but trust in science is far more complex than what seems to be suggested by such a definition (Funk et al., 2019). Many contemporary skeptics and denialists critique science today for its lack of certainty and its reliance on speculative mathematical theories, even though these are cast as reliable plausible reasoning and high-ranking epistemic warrants (Pittinsky, 2015).

Faulkner and Simpson, in their 2017 book *The Philosophy of Trust,* emphasize that trust involves *dependence relations* and an *expectation* that the trusted will see this dependence *as a reason* to act according to the expectation. In other words, there is an expectation on behalf of the trusting agent that the other will find it reasonable to engage in a trusting relationship. So, questions regarding science, such as "should we trust a theory" or "should we trust the mathematics, and therefore not worry about empirically testing the theory?" need to be considered not only in terms of epistemic reliability, but also in terms of the dependency relations and affective bonds and aesthetic investments that strongly constitute feelings of expectation. The philosopher Stern (2017) suggests that trust is not so much of our making as it is given in life itself, a public good that we cannot take credit for, in a world where affect and feelings of betrayal both fuel and destabilize our collaborations. This less pragmatic assessment treats trust as beyond discernment; trust becomes basic to life in some fundamental way, being a capacity or tendency that belongs first and foremost to a relational ecology.

When trust is pronounced in a situation, and the stakes are high, feelings can move from disappointment to betrayal when that trust is broken. Latour (2018) suggests that the widespread "epistemological delirium" that characterizes denial of climate science must be understood as an effect of betrayal, in that people have been denied the globe that was promised in the name of Modernism:

> … this "people" has been coldly betrayed by those who have given up the idea of actually pursuing the modernization of the planet with everyone, because they [the elites] knew, before everyone else, that such modernization was impossible – precisely for want of a planet vast enough for their dreams of growth for all. Before accusing "the people" of no longer believing in anything, one ought to measure the effect of that overwhelming betrayal on people's level of trust. Trust has been abandoned along the wayside.
>
> *(pp. 22–23)*

And yet Rutjens et al. (2017) document how contemporary distrust in science in the US has emerged in different ways, and that the climate is not the only trouble point. The five key topics that elicit distrust are climate science, stem cell research, evolution, the Big Bang Theory, and nano-technology (Blake, 2015). Although these five all draw from different reasons for that distrust, there are marked correlations. For instance, studies show that endorsement of a free-market ideology can reliably predict skepticism in the anthropogenic causes of climate change (Lewandowsky & Oberauer, 2016; Lewandowsky et al, 2013). The Union of Concerned Scientists and various speakers at the March for Science have declared that Christian right organizations are at war with science. Other political leaders interested only in free market economics line up beside these organizations (Mooney, 2005). However, a *Nature* editorial suggested that casting the situation as a "war on science" was too polarizing (Nature, 2017). I agree that there are bigger sociotechnical changes at play, concerning first a massive undermining of public education, and a president who prizes entrepreneurs over experts, and claims to possess "a natural instinct for science."

Education plays a complex role in science dis/trust. Studies show that further education does not necessarily alter beliefs about the trustworthiness of science (Drummond & Fischhoff, 2017). These studies found that taking a science course at college, perhaps as a mandatory course, does not necessarily cause students to embrace scientific knowledge, but might simply reinforce their initial suspicions. Ironically, these insights are based on data from national surveys whose questions actually entrench narrow images of science. In other words, multiple-choice survey questions forced respondents to affirm or deny their trust in science, without soliciting a more nuanced evaluation that might explore degrees of trust, the complex webbing of trust, and their image of science.

In response to the current denialism of climate science, the US National Science Teachers Association published an official statement, offering support and guidance on how to teach climate science (NSTA, 2018). Notably, the document points to the "overwhelming scientific consensus" to support its position, declaring: "The scientific consensus on the occurrence, causes, and consequences of climate change is both broad and deep" (Melillo et al., 2014). Claims to consensus are crucial in both establishing scientific truth within the community of scientists but also in contesting the authority of science within the community of skeptics. Denialists are often seeking absolute truths and are thus disappointed by, and indeed distrustful of, the consensus-making habits of scientists (Oreskes, 2019).

Latour (2018) recognizes that some of his own scholarship might have been misconstrued as anti-science. Scholarship in Science and Technology Studies (STS) has shown how scientists, like the rest of us, are opportunistic and biased, and often working in the service of sexism, racism, and colonialism. As Yusoff (2018) explains, geological sciences, for example, have contributed to the climate crisis through their complicit partnership with corporate mining companies. For this and other reasons, she argues that there is no one humanity that suffers the Anthropocene, but a multiple earth of a billion black anthropocenes (Yusoff, 2018). Tsing (2015) and Haraway and many others turn to terms like Capitalocene, Anthrobcene, Plantationocene to bring this point home; important links between colonialism and climate make evident this impact (Chakrabarty, 2009, 2012, 2014). And yet to contest the "we" of the Anthropocene altogether can, as we see with contemporary science denialists, further fuel a free market capitalism which benefits from continued inequity and continued corporate acquisition (Danowski & Viveiros de Castro, 2017). The term "Climatic regime" reminds us that "the climate" is never a pure state of nature operating outside of culture, but always populated and engendered by diverse agencies, some of them human. I have explored these tensions in more detail elsewhere (de Freitas, 2019; de Freitas & Weaver, 2019).

Some new materialist theories advocate for an image of science as an activity that entails forming alliances with nonhuman agencies. Latour (2017, 2018) argues that science is one of our best pluralistic "engendering" practices to be used as we investigate the complex agentic "metamorphic zone" of the planet he calls "Gaia" (see Chapter 2). If science is at its best when proliferating non-human agencies and remixing the metamorphic zone, how might it guard against a sentimentalism or biopolitics of feeling that imprints whiteness or humanity onto everything that was once alien or other? How should we trust science in a relational ontology characterized by this radical contingency and more-than-human agency? Since hierarchical power relations persist, a "healthy skepticism" toward scientific claims must be considered part of a cosmic context (Stengers, 2018). Within such posthuman ecologies, we are called to keep front and center the political nature of any pluralist science of dependency relations and "engendering" practices. And since speculative thought is the power horse behind such practices, we need to look a little closer at how speculation works in science. In particular, I want to grapple with the links between speculation, abstraction, and model. Mathematical models have been the very instruments for furthering our understanding of climate change, and yet they are also global-scale abstractions typically dislocated from the local. This dilemma concerns the distinctive capacities of mathematics, as both an extractive force and a speculative invention, an act of abstraction that can either be reductive or creative, serving to both mis/represent the real and yet also to imagine otherwise.

Mathematical modeling

Wark (2017) suspects the new materialism of Bruno Latour is simply not well equipped to deal with big picture thinking about climate science, nor with the powerful role that mathematical models and simulations play in climate science: "no amount of agential complexity or empirical complication really disturbs that big picture" (Wark, 2017, np). The bottom-up agential realism of Latour and Barad tends to shy away from the kind of structural abstractions we mobilize in climate models. Climate science models may be based in big data techniques, emerging from a constant feedback loop and stochastic training on massive amounts of data points, but they are packaged as large-scale abstract models that take on a life of their own. Wark seems to think that Latour has banished the art of abstraction – and banished mathematics as bad abstraction. We can see this tendency in Latour's work, as he often casts mathematical thinking as a reduction of complexity and a betrayal of immanence. But I think Latour is mistaken to blame mathematics for serving the "view from nowhere" in part because, in doing so, he is sacrificing the speculative power of abstract thought. Indeed, I suspect we will need to think more creatively about mathematical abstraction – as a generative activity (and not only as a reductive subtraction) and as a creative act that brings the new forth, making it part of the metamorphic zone where truth is contested. To be fair, Latour has written insightfully about the nature of the quantitative, inspired by the early posthumanist sociologist Gabriele Tarde, and shown how new tools in the digital humanities could be used to study complex science networks (see Chapter 3). Mathematical abstraction, however, is far more robust and diverse than what we find in big-data network analysis.

There are certain relational ontologies that do incorporate mathematics, but they often ascribe, problematically, to a "naturalized metaphysics" and "eliminative materialism" or "physicalism." These are typically *scientistic* in what is often called a form of "strong realism" and committed to the ultimate unity of science, and in some cases to an all-encompassing theory of everything (ToE). For instance, James Ladyman and Don Ross (2007), in their book *Everything must go*, argue for a "relational metaphysics," which rests on homotype theory (a kind of set theory/logic), as the ontological foundation of the real. They ascribe to a total identity between physical and mathematical structures, calling the approach "structural realism." Their work presents the quintessential example of reductive formalisms. They are adamantly scientistic in their claim that "scientific institutional processes are absolutely and exclusively authoritative" (Ladyman & Ross, 2007, p. 65). They suggest that their realism escapes the trap of Kantian correlationism, in which the real is said to only *correlate* to human descriptions and knowledge of it. Realisms are an antidote to correlationism. Their particular structural realism claims that science can come to know reality *as it is independent of us*. I am interested in realisms that attend to dependency rather than independence. The notion of independence is at the heart of many disagreements between new materialists and realists, including speculative realists like Graham Harman (see Chapter 4). It is for this reason that I dwell on the notion of *dependency relations* within the metamorphic zone, or rather, on how dependency relations are what characterizes absolute contingency. In other words, we might typically associate a random collection of free *independent* agents with contingency, because of their unpredictable and uncorrelated nature, but that would be to misrecognize the role of chance. Contingency is poorly conceived in the throw of the dice. Contingency thrives only when we affirm "all of chance" which can only be done through reckoning and engaging with the complex web of dependency relations that form a relational ontology (more on this below).

In terms of an image of science, foundational projects, like that of Ladyman and Ross, are dangerous erasures of the plurality of practice, in attempting to buttress epistemic truth-preserving procedures. Yes, they commit to a kind of relational ontology and ontological realism, much like many new materialists, but their image of science is thoroughly positivistic. Thinkers like Latour and Stengers might claim that Ladyman and Ross misrecognize science, because (1) they fail to incorporate the efforts of science to *fabulate* and speculate and (2) they fail to really reckon with the more-than-human agentic powers immanent within a *metamorphic* nature. For new materialists, being a part of the metamorphic zone is actually how science – and I would argue mathematics – effectively procures knowledge of the real. Ian James (2019), who attempts to walk a zig-zag path through all these thinkers, suggests that scientific realisms like that of Ladyman and Ross need a dose of insufficiency, modesty, and sympathy.

Mathematics plays an enigmatic role in debates regarding the authority of science. Scientific theories that seem to rely excessively on mathematical prowess are often considered overly speculative. At the 2015 Munich conference *Why trust a theory? Reconsidering scientific methodology in light of modern physics,* Kragh notes that "the classical demarcation problem between science and non-science has taken a new turn with the appearance of fundamental and highly mathematical theories which may not be experimentally testable in the ordinary sense" (n.p). According to the physicist Hossenfelder (2018) in her book *Lost in Math: How Physics Is Led Astray by Beauty*, physics has lost touch with the real because superstring theories and multiverse hypotheses are so caught up with mathematical objectives of naturalness and aesthetic balance. She suggests that physics is no longer physics but some kind of pure mathematical speculation, a kind of "post-empirical physics," which is not to be trusted as a reliable description of the world. The philosopher Shapere (2000, pp. 159–161) states rather controversially that "physics is in fact approaching, or perhaps has reached, the stage where we can proceed without the need to subject our further theories to empirical test." But Kragh (2015) and others point out that non-empirical theory building in physics is not new.

Speculative materialism

I turn now to the SF novel *The Three-Body Problem* precisely because such stories take up the speculative limits of science, probing the potential of mathematics as a creative worlding process. As Haraway (2013) contends, SF novels activate an "opening up what is yet-to-come in protean entangled times' pasts, presents, and futures" (n.p). In SF, the present historical moment, with its particular sociotechnical aspect, is used as a base for extreme extrapolation, producing another world that is weirdly and shockingly possible. For Haraway, this "opening up" is always political and charges those who speculate with a responsibility for the worlds they bring into being, and for those who will "bear the consequences" of the created world (Truman, 2019). Freedman (2000) notes that science fiction as a genre is "the one most devoted to the historical concreteness and the historical self-reflectiveness of critical theory" (Freedman, 2000, p.xvi). In science fiction, scientific concepts and practices are often queered or extrapolated to extremes using thought experiments and narrative devices and tropes and can help readers imagine a radically different form of empiricism. SF stories can put flesh on a thought experiment and do philosophy through character and story. These speculations function politically, exploring the limits of science, while tracing the links between science and society (Shaviro, 2015).

The novel *The Three-Body Problem* sheds considerable light on contemporary issues of dis/trust in science and mathematical modeling within a chaotic climate (Li & Isaacson, 2019). The novel follows the principal character Wang Miao, a nano-technology engineer, as he learns about an alien Trisolaran world subject to a chaotic climatic regime. The Trisolarans must solve a physico-mathematical problem called "the three-body problem" in order to save their odd solar system and predict the movement of their three suns, which move chaotically across their skies. The Trisolaran world lacks habitable regularity and seems to undermine the very possibility of science. The only invariants and scientific claims possible in this world are empty tautologies: "Other than Stable Eras, all times are Chaotic Eras" (Liu, 2014, p. 96). Such statements have no conventional content value because they offer no way forward for the conventional scientific method. There seems to be no pattern of solar behavior that is invariant in the Trisolaran world, and therefore no possibility for science. How can science work in such a world of absolute contingency? In *The Three-Body Problem*, Trisolarans are nonetheless capable of science-as-war, and stealthily attack the Earth from afar, sending modified protons – named "sophons" – to interfere at the imperceptible scale in scientific experiments on Earth. Sophons undermine the replicability of the experimental method on Earth, destroy human science, and leave the Earth vulnerable to invasion. The Trisolaran aim is to undermine trust in science on Earth, using various tactics, until scientists come to conclude that "physics does not exist."

The gravitational problem of three-bodies is a classical problem that has resisted solution over the centuries. The problem was of huge industrial importance in the 18th century, because of its relevance to navigation and European colonial exploits. The French mathematician Henri Poincaré (1854–1912), known also for his work on global telecommunications, showed that the problem cannot be resolved using classical analysis due to the sensitivity of the system to initial conditions. In other words, there is no closed-form solution (captured in a parametric model), and we must use brute-force methods to capture the unique systemic relations in any given three-body situation. The problem has hence been considered unsolvable in the general case. There has been progress in the last decades in what are called "periodic" solutions using multiple computational simulations (Liao & Li, 2019). In the fictional context of *The Three-Body Problem*, the main character Wang participates in various attempts to solve the problem. At one point, the character Wei Cheng, a solipsistic mathematician, develops an "evolutionary algorithm" that promises to solve the problem. The reference to an evolutionary algorithm evokes genetic and generative computing, where the search engine mobilizes a kind of procedural randomness, like that encoded in machine learning. These algorithms mimic some aspects of biological evolution. Sadly for the Trisolarans, this computational big-data approach, relying on extractive data mining techniques rather than analytic models, also fails to help. The attempt to use brute computational methods – which are enacted in the simulated game-world involving vast armies of binary agents – is simply not adequate to the challenge of this truly chaotic system.

In a short essay called *Science Fiction and Extro-Science Fiction*, Meillassoux (2015) distinguishes between two kinds of fiction, SF and XSF. The first is the more familiar, SF, and stands for science fiction, while XSF stands for a genre of writing he calls extro-science-fiction. SF always entails a "fictional future of science that modifies, and often expands, its possibilities of knowledge and mastery of the real" (pp. 4–5) while XSF (or in French, *fiction hors-science*) is a fiction of "worlds outside-science" (p. 5). These latter worlds are not simply without science but are rather worlds where science is impossible. XSF is an attempt to think of a world outside science, a world that cannot be configured as an object of scientific inquiry, that is, a world that is, in principle or in essence, inaccessible to science. Meillassoux finds that most SF fiction cannot achieve this. SF stories are always opening up the possibility of science to expand or grow, a science that stretches the limits of our current practice. He seeks a story that breaks science, and its presumed correlation to the real, a story that can imagine a radical outside to the relational ontology in which our science effectively works. One wonders whether the engine of story itself imposes structural constraints on alt-worlding so strongly that "laws of nature" always emerge, and hence a kind of science, as we've come to know it. For Meillassoux, a lawless world would be without science, though not without other capacities. In other words, it would *not* be utter mayhem. There could be another kind of regularity, and, I would imagine, another kind of trustworthy method that achieves some sort of more-than-human understanding of the world, achieved without reliance on classical or probabilistic science. The existence of such a fiction (XSF) would break with Kantian co-relationism, so as to posit a radical outside or non-relational, which Meillassoux describes as a world ("our" world inasmuch as we partake) of *necessary and absolute contingency*. His project is to pursue a kind of speculative materialism that escapes the correlationism that has dominated post-Kantian philosophy. In other words, he seeks more than "historical" contingency, whereby the science of one world might in principle mutate into the science of another world.

When turning to speculative fiction in search of examples where there is a robust attempt to disassociate the conditions of science from the conditions of thought, we expect to find some form of inquiry incommensurable with our current scientific practice. According to Meillassoux, it would be a world where irregularity would be sufficient to abolish science but not consciousness: "A world whose margins would become capricious, but this caprice would not refer to any hidden intention. A world in which we could only chronicle things" (Meillassoux, 2015, p. 36). This raises the question: What forms of inquiry will do justice to the trickster nature of matter? (Carstens, 2017). Would there only be one legitimate method in such a world, a method that simply aimed to chronicle events? The disorder of such a world would be so manifest that any of those who persisted in imagining a hidden law to explain the absurd variations of nature would seem "eccentric or vain" (Meillassoux, 2015, p. 37). In the earlier and fabulous essay *After Finitude*, Meillassoux (2008) turns to mathematics in search of a radical ontology beyond the correlational. For Meillasoux (2008), the direct mathematization of science and the natural world, in the work of early moderns like Galileo, was too radical and too alienating for European science, which timidly retreated to the Kantian screen of representation. The Moderns inherited a timid or tentative relationship with nature (Latour, 2015). According to Meillassoux, Kantian correlationism keeps the real always veiled behind the conceits of language and human faculties. As such, mathematics is only a system of signs that *represents* the real, which is, lucky for us, *a priori* structured to correlate with human cognition. Correlationism disallows an encounter with the radical outside, a regime of truth that Deleuze and Guattari will seek to overthrow. Meillassoux focuses especially on the mathematics of Cantor's transfinite numbers, where different countable and uncountable infinities are compared. He suggests that Cantor's diagonal counting of the permuted rational numbers (1891) was revolutionary because it affirmed an incommensurable outside. The transfinite uncountables became the outside, although the act of affirming their status was simultaneously the act of packaging them into set theory. Pourciau (2022) makes a strong case for a different reading of Cantor's contribution and interprets his transfinite revolution as the first nail in the digital coffin. She draws a direct link between Cantor and Turing and suggests that Cantor's set theory is the foundation of discrete state machines. Ironically, mathematics gets us closer to absolute contingency than most other endeavors, while keeping us shackled to discrete structures.

Is *The Three-Body Problem* an example of XSF? Does it mobilize math in that way? Does it "present" an outside or alien world, replete with disorder and disunity, into which the human world is thrown, where the inherent contingency of law-like behaviour is exposed as necessary and absolute? Or does the novel simply pursue a Humanist image of science, enacting a colonial intergalactic acquisition? These questions matter philosophically, as we ponder the ways in which *relationality* is a self-serving cosmic grasping gesture, an attempt to be at home everywhere. In exploring an outside to science, we are also exploring an outside to a particular kind of relationality. Initially, I had thought that *The Three-Body Problem* was a good candidate for XSF, but the more I considered both the book and the category of XSF, the more the story seemed an exemplary work of SF, and yet, simultaneously, the more I became confused about the very nature of XSF. The novel presents a world where the laws of nature are radically altered, and where ultimately "physics does not exist"; the Trisolaran intervention of crafting and disseminating destructive "sophon" particles (sophon being an interesting mixture of wisdom and matter) truly undermines science. However, the novel retreats to a humanist image of worlding, and a human all too human image of reason. The science in the novel reflects historical context (i.e. the Chinese Cultural Revolution). Despite reframing scientific practice as more-than-human and indifferent to human survival, it seems in the end that the scientific advancements – both the science performed by the Trisolarans and scientific advancements on Earth – are entirely fueled by imperialism. Much of the plot celebrates human exceptionalism and continues to invest in "causalist" interpretations of scientific intervention.

Concluding gesture: all of chance

Trust in science has been sustained by a certain kind of "belief in the world," to borrow Deleuze's (1985) refrain. That belief points to a default realism and presumption that the world is there for us, awaiting description and experiment. That belief in the world has also granted science its authority, and often secured our trust in science. But Deleuze suggests that a belief in the world has to be humble – the world might not be providentially available. We could learn so much from thinking of contingency in terms of trickster worlds, which are playful and often indifferent to human ambition (Burkhart, 2019). Science denialists cling to their belief in the world, while others surrender belief to a new image of science that rests instead on incremental cybernetic feedback loops. The latter folks live in what Châtelet (2014) calls a "thermocracy" where humans act like billiard balls, their collisions accumulating into subject positions and belief systems. Why do neither of these options satisfy or make for a truly relational ethico-onto-epistemology? Because neither are the kind of belief in the world we need, and neither convey what

Deleuze envisioned. Belief in the world must entail a speculative openness to that which is not commensurable or relatable. Paradoxically, such belief is to live "in relation" to an outside of relationality and to appreciate the fundamental "absolute" contingency of life. A belief in the world has to open onto that contingency, not as automated randomness in an evolutionary algorithm, which fails to do "chance" justice. But nor in an attempt to shut down difference and discomfort, by hiding behind "alternative truths." A belief in the world, for Deleuze, must learn to affirm "all of chance" and not simply invest or trust in the predictive weighted outcomes of some probabilistic measure (Deleuze, 1990, p. 60). For a form of inquiry to affirm "all of chance," it would have to stretch into the actual infinite of the speculative imaginary and engage with absolute contingency. Inquiry under these bizarre conditions could not be packaged and sold as a unitary epistemological encoding of passive unified matter. It would splinter into a thousand practices, and there would be a fundamental disunity to inquiry methods. Belief and trust would then be grounded in an affirmation of a multiple earth, demanding multiple "sciences," each enthralled to a paradoxical embrace of *all of chance*. As I ponder this strange world, I realize this entails accepting that science is not unified around truths, but around trusted rituals of engagement in risky forms of diplomacy.

Speculative fictions like *The Three-Body Problem* help us to grasp the deeply complex nature of empirical inquiry – a practice that typically involves colonial conquest, diplomacy, betrayal, and pluralist ontogenerative interventions. If future empiricisms are to expand alliances across posthuman ecologies, while still invested in imperial world building, we must maintain a skepticism and even a certain capacity for betrayal of the ideals or disciplines that established the conditions for the inquiry. Surely, the ethics of such future practices will involve responding to the emergent dependency relations in the metamorphic zone, so that forms of inquiry will expand sensitivities and increase our comprehension of the abstract work of chance and contingency. In living with the possibility of such betrayal, and "staying with the trouble" as Haraway (2016) reminds us, science as a speculative activity has the potential to proliferate agencies in a pluralistic and potentially more inclusive future. The plot and poetics of speculative fiction novels like *The Three-Body Problem* immerse us as readers into strange and estranged relational ontologies. SF stories force us to grapple with the outside of science. And they show the power of speculative thought – as promiscuous venture – transgressing the borders of mathematical necessity and sufficiency, toward the necessity of contingency.

References

Alumkal, A. (2017). *Paranoid science: The Christian right's war on reality*. New York: New York University Press.

Åsberg, C., Thiele, K., & van der Tuin, I. (2015). Speculative before the turn: Reintroducing feminist materialist performativity. *Cultural Studies Review*, 21(2), 145–172.

Bahng, A. (2017). Plasmoidial improprieties: Octavia E. Butler, Slime molds, and imagining a Femi-Queer commons. In C. Cipolla, K. Gupta, D. A. Rubin, & A. Willey (Eds.), *Queer feminist science studies* (pp. 310–326). Seattle: University of Washington Press.

Barnes, E. (1997). *States of sympathy: Seduction and democracy in the American novel*. Columbia University Press.

Blake, A. (2015). Americans increasing distrust in science – and not just in climate change. *The Washington Post*. 30 January. Available at https://www.washingtonpost.com/news/the-fix/wp/

2015/01/30/americans-increasing-distrust-of-science-and-not-just-on-climate-change/ (accessed 21 September 2020).
Burkhart, B. (2019). *Indigenizing philosophy through the land: A trickster methodology for decolonizing environmental ethics and Indigenous futures*. Michigan State University Press.
Carstens, D. (2017). Tricksters, animals, new materialities and indigenous wisdoms. In W. Woodward, & S. McHugh (Eds.), *Indigenous creatures, native knowledges, and the arts: Animal studies in modern worlds* (pp. 93–115). Cham: Palgrave.
Chakrabarty, D. (2009). The climate of history: Four theses. *Critical Inquiry*, 35(2), 197–222.
Chakrabarty, D. (2012). Postcolonial studies and the challenge of climate change. *New Literary History*, 43(1), 1–18.
Chakrabarty, D. (2014). Climate and capitalism: On conjoined histories. *Critical Inquiry*, 42(1), 1–23.
Châtelet, G. (2014). *To live like pigs: The incitement of envy and boredom in market democracies*. (Trans. R. Mackay). New York: Urbanomics.
Colebrook, C. (2019a). A cut in relationality: Art at the end of the world. *Angelaki: Journal of the Theoretical Humanities*, 24(3), 175–195.
Colebrook, C. (2019b). *Unthinkable extinction: Cinematic time and the panorama of history. Alienocene: Journal of the First Outernation*. Available at https://alienocene.com/2019/06/19/unthinkable-extinction/ (accessed 23 September 2020).
Danowski, D., & Viveiros de Castro, E. (2017). *The ends of the world*. Cambridge: Polity.
de Freitas, E. (2017). Non-human findings from the laboratory of speculative sociology. *Minnesota Review*, 88, 116–126.
de Freitas, E. (2019). *Bruno Latour and distrust in science*. Lecture given to The Royal Society for Philosophy, Human Sciences Seminar Series. Manchester Metropolitan University, February.
de Freitas, E. (2020). Science studies and the metamorphic multiple earth: Risky diplomacies in the end-times. *Cultural Studies ↔ Critical Methodologies*, 20(3), 203–212.
de Freitas, E. (2023). The role of abduction in mathematics: Creativity, contingency, and constraint. In L. Magnani (Ed.), *Handbook of abductive cognition*. Springer Verlag. https://link.springer.com/referenceworkentry/10.1007/978-3-030-68436-5_34-1
de Freitas, E. (2024). *Xenomathematics. Lecture in proceedings*. Quebec City: Canadian Mathematics Education Study Group.
de Freitas, E. (2025a). Modeling and worlding: How to imagine more-than-human mathematics. In K. Le Roux, A. Coles, A. Solares-Rojas, A. Bose, C. Vistro-Yu, et al. (Eds.), *Proceedings of the 27th ICMI Study Conference (Mathematics Education and the Socio-Ecological)* (pp. 76–82). MATHTED and ICMI. permanent url: https://www.mathunion.org/icmi/icmi-study-27
de Freitas, E. (2025b). Xenomathematics as a planetary practice: Imagining more-than-human mathematical worlds. *Educational Studies in Mathematics*. DOI:10.1007/s10649-025-10424-x
de Freitas, E., & Truman, S. (2020a). New empiricisms in the Anthropocene: Thinking with speculative fiction about science and social inquiry. *Qualitative Inquiry*. https://doi.org/10.1177/1077800420943643
de Freitas, E., & Truman, S. (2020b). Science fiction and science dis/trust: Thinking with Bruno Latour's Gaia and Liu Cixin's The three body problem. *Rhizomes: Cultural Studies in Emerging Knowledge*. Available at http://www.rhizomes.net/issue36/defreitas-truman.html (accessed 11 April 2025).
de Freitas, E., & Weaver, J. (2020). Rethinking social inquiry in the wake of science studies: Transdiscipline pursuits in times of climate change, information flows, and fading democracies. *Cultural Studies – Critical Methodologies*, 20(3), 195–202.
Deleuze, G. (1985). *Cinema II: The time image*. (Trans. H. Tomlinson & R. Galeta). London: Bloomsbury Press.
Deleuze, G. (1990). *Logic of sense*. (Trans. M. Lester & C. Stivale). New York: Columbia University Press.
Drummond, C., & Fischhoff, B. (2017). Individuals with greater science literacy and education have more polarized beliefs on controversial science topics. *Proceedings of the National Academy of Sciences*, 114(36), 9587–9592.

Dupré, J. (2003). *The disorder of things: The metaphysical foundations of the disunity of science*. Cambridge, MA: Harvard University Press.

Faulkner, P., & Simpson, T. (2017). *The philosophy of trust*. Oxford: Oxford University Press.

Freedman, C. (2000). *Critical theory and science fiction*. Hanover: Wesleyan University Press.

Funk, C., Hefferon, M., Kennedy, B., & Johnson, C. (2019). *Trust and mistrust in Americans views of scientific experts*. Pew Research Center: Science and Society. Available at https://www.pewresearch.org/science/2019/08/02/trust-and-mistrust-in-americans-views-of-scientific-experts/ (accessed 21 September 2020).

Gauchat, G. (2012). Politicization of science in the public sphere. A study of public trust in the United States, 1974 to 2010. *American Sociological Review*, 77(2), 167–187.

Hansson, S. (2017). Science denial as a form of pseudoscience. *Studies in History and Philosophy of Science Part A*, 63, 39–47.

Haraway, D. (2013). SF: Science fiction, speculative fabulation, string figures, so far. *Ada: A Journal of Gender, New Media, and Technology*, 3. Available at http://www.markfoster.net/dcf/speculative_fabulation.pdf (accessed 21 August 2025).

Haraway, D. (2016). *Staying with the trouble: Making Kin in the Chthulucene*. Durham, NC: Duke University Press.

Hossenfelder, S. (2018). *Lost in math: How beauty leads physics astray*. NY: Hachett Book Group.

James, I. (2019). *The technique of thought: Nancy, Laruelle, Malabou and Stenger after naturalism*. University of Minnesota Press.

Keeling, K. (2019). *Queer times, black futures*. New York: New York University Press.

Kragh, H. (2015). Fundamental theories and epistemic shifts: Can history of science serve as a guide? Paper presented at the conference: Why trust a theory? Reconsidering scientific methodology in light of modern physics. Available at https://videoonline.edu.lmu.de/en/node/7481 (accessed 23 September 2020).

Ladyman, J., & Ross, D. (2007). *Every thing must go*. Oxford: Oxford University Press.

Latour, B. (2015). Latour in conversation with Heather Davis. In H. Davis, & E. Turpin (Eds.), *Art in the Anthropocene: Encounters among aesthetics, politics, environments and epistemologies* (pp. 43–65). London: Open Humanities Press.

Latour, B. (2017). *Facing Gaia: Eight lectures on the new climatic regime*. Oxford: Polity Press.

Latour, B. (2018). *Down to earth*. Oxford: Polity Press.

Lewandowsky, S., Gignac, G. E., & Oberauer, K. (2013). The role of conspiracist ideation and worldviews in predicting rejection of science. *PLoS One*, 8(10), e75637.

Lewandowsky, S., & Oberauer, K. (2016). Motivated rejection of science. *Current Directions in Psychological Science*, 25, 217–222.

Li, G., & Isaacson, N. (2019). China turns outward: On the literary significance of Liu Cixin's science fiction. *Science-Fiction Studies*, 46(1(137)), 1–20.

Liao, S., & Li, X. (2019). On the periodic solutions of the three-body problem. *National Science Review*, 6(6), 1070–1071.

Liu, C. (2014). *The three-body problem*. New York: Tor.

Lothian, A. (2018). *Old futures: Speculative fiction and queer possibility*. New York: New York University Press.

Meillassoux, Q. (2015). *Science fiction and extro-science fiction*. Minneapolis: University of Minnesota Press.

Meillassoux, Q. (2008). *After Finitude: An Essay on the Necessity of Contingency*, New York: Continuum Press.

Melillo, J. M., Richmond, T.C., & Yohe, G. (Eds.) (2014). *Climate change impacts in the United States: The third national climate assessment*. Washington, DC: U.S. Global Change Research Program.

Mooney, C. (2005). *The Republican war on science*. New York: Basic Books.

National Science Teaching Association (NSTA). (2018). *NSTA position statement on the teaching of climate science*. Arlington, VA: National Science Teaching Association.

Nature Editorial. (2017). Beware the anti-science label. *Nature*, 545, 133–134.

Oreskes, N. (2019). *Why trust science?* Princeton: Princeton University Press.

Pittinsky, T. (2015). America's crisis of faith in science. *Science*, 348, 511–512.

Pourciau, S. (2022). On the digital ocean. *Critical Inquiry*, 48(2), 233–261.

Rieder, J. (2008). *Colonialism and the emergence of science fiction*. Middletown, CT: Wesleyan University Press.

Rutjens, B., Sutton, R., & van der Lee, R. (2017). Not all skepticism is equal: Exploring the ideological antecedents of science acceptance and rejection. *Personality and Social Psychology Bulletin*, 44(3), 384–405.

Schuller, K. (2018). *The biopolitics of feeling: Race, sex and science in the 19th century*. Durham, NC: Duke University Press.

Shapere, D. (2000). Testability and empiricism. In *The reality of the unobservable: Observability, Unobservability, and their impact on the issue of scientific realism* (pp. 153–164). Berlin: SpringerNature.

Shaviro, S. (2015). *Discognition*. London: Repeater.

Stengers, I. (2018). *Another science is possible: A manifesto for slow science*. Cambridge: Polity Press.

Stern, R. (2017). Trust is basic: Logstrup on the priority of trust. In P. Faulkner & T. Simpson (Eds.), *The philosophy of trust* (pp. 272–294). Oxford: Oxford University Press.

Truman, S. E. (2019). SF! Haraway's situated feminisms and speculative fabulations in English class. *Studies in Philosophy and Education*, 38(1), 31–42.

Tsing, A. (2015). *The mushroom at the end of the world: On the possibility of life in capitalist ruins*. Princeton, NJ: Princeton University Press.

Wark, M. (2017). Bruno Latour: Occupy earth. Verso Blog, 5 October. Available at https://www.versobooks.com/blogs/3425-bruno-latour-occupy-earth (accessed 23 September 2020).

Yusoff, K. (2018). *A billion black anthropocenes or none*. Minneapolis, MN: University of Minnesota Press.

12

COMPUTATIONAL THINKING AND THE INFANT MIND

Software studies between empiricism and nativism

Elizabeth de Freitas and Matthew Curinga

Nativist versus empiricist epistemology

Classical epistemological debates between nativism and empiricism carry over into discussions about artificial intelligence. Advocates of deep neural net learning (DL) are usually cast as empiricists and advocates of more traditional symbolic AI models are cast as nativists, with the caveat that such glossing of this debate overlooks a great deal of nuance, but helps us map the theoretical stakes of our current machine learning models. In this chapter, we will use these terms – empiricist and nativist – as a means of making sense of some of the recent developments in computational paradigms, keeping this caveat in mind. The nativists often critique deep learning AI by citing studies of child behaviour and the capacity of infants to learn quickly after only one or few encounters ("one-shot agile learning") to support the claim that *a priori* cognitive symbolic structures are the foundation of human and machine intelligence. Despite the impressive ability of large language models (i.e. Chat GPT) to simulate human language-use, many thinkers criticize these amped-up models because they are *inefficient learners* when compared to humans, requiring enormous amounts of data and energy, and forming rigid concepts that lack the cognitive suppleness and adaptive learning one finds in children (Lake et al., 2015; Levine, 2023). Deep neural net AI models are thus castigated as *poor learners* because of the excessive amount of data they require, and because they cannot pivot easily after learning from mistakes. Noam Chomsky and Gary Marcus, as well as computer scientists like François Chollet, Branden Lake, and Josh Tanenbaum, refer to studies of childhood developmental psychology to support their preference for a machine intelligence that would supplement neural nets with more traditional symbolic AI – an approach often referred to as "neuro-symbolic" (see Ellis et al., 2020; Lake et al., 2015). Indeed, Chollet (2019) has offered a challenge to the computer science community to find a rapprochement between deep learning and symbolic approaches. Notably, as evidence of this retreat to previous cognitive models of innate childhood learning, Chollet takes up the theories of cognitive psychologist Elizabeth Spelke at the Harvard *Center for Brains, Minds and Machines*, who claims that the aim of AI should be to "reverse engineer the infant mind and create smart machines" (Spelke, 2020).

DOI: 10.4324/9781003538769-12

In this chapter, we show how the debate between neural and symbolic AI rests on a particular image of the infant mind, elaborated by Seymour Papert and Marvin Minsky at the MIT Artificial Intelligence Lab in the 1960s. We then track how this image of the child's mind was displaced within cognitive science by the concept of "computational thinking" (CT) during the 2000s, when neural network approaches to AI began to dominate the field. CT was meant to characterize *expert* reasoning in STEM, rather than children's reasoning, and recast the nature of machine/human abstraction in terms of pattern recognition. We then turn to current work in Software Studies, which we characterize in terms of a post-Kantian empiricism– specifically the investigative aesthetics of Matthew Fuller & Eyal Weizman, and the network media analysis of Alexander Galloway. We call this new empiricism "post-Kantian" to help situate it within the debates between nativism and empiricism mentioned above and to underscore the *critical* dimension of such approaches. Our aim in this chapter is to tease out some of these different images of cognition and intelligence, as part of a shift in the epistemological underpinnings of our current algorithmic condition. Galloway (2021) suggests that the "network" image of intelligence has replaced all earlier ideas about knowledge, learning, and computing, and links this to behaviourism's relinquishing of interiority over the last decades. Empiricism - in the form of behaviourism - underpins deep learning models because they are effectively an input-output loop, without interior built-in programs. This new image distributes thought across an information connectionist network and turns away from prior onto-epistemic assumptions of innate cognitive structure. Minsky and Papert's Kantian-inspired children are no more. Obfuscation moves from within (the cloaked innate capacity of the human mind) to the vast interleaved hypothesis space of opaquely layered neural nets.

The Kantian influence of Jean Piaget

One of the founders of AI, Seymour Papert (1928–2015), worked with Jean Piaget (1896–1980) at the University of Geneva, between 1958 and 1962, and was one of Piaget's protégés, before co-directing the Artificial Intelligence Lab at MIT. Piaget's theories of childhood development and "genetic epistemology" strongly influenced the newly emerging cognitive science of the 1960s and the burgeoning field of symbolic AI. At the heart of Piaget's cognitive psychological approach was both an experimental methodology and a developmental ideology that stipulated the order and age at which children (can/should) develop specific conceptual skills (Piaget, 1954). This approach has been hugely influential in theories of learning. Papert brought Piaget's experimental methodologies and ideas about learning to his work in the AI lab at MIT. Together with Marvin Minsky, Papert published the 1972 *Progress Report on Artificial Intelligence*, which explained how their studies had "become closely bound to the study of development of intelligence in children" (p. 2). In their report, Minsky and Papert critique the early artificial neural nets (i.e. early connectionist efforts in the neural network tradition) of the 1950s–1960s, noting that "the descriptive powers of these quasi-linear learning schemes have such peculiar and crippling limitations that they can be used only in special ways" (p. 31). In other words, they strongly dismissed the deep learning models and claimed that cranking through empirical data could only support limited kinds of learning, and what was needed was a more nativist approach (a symbolic AI model). The criticism was quite harsh, although they admitted that sometimes a "weighted decision" and "incremental adaptation" are better than nothing. Still, and foreshadowing what would come, they worried that simply layering neural nets only masks the problem of what they called the "terminal" learning trajectory of neural nets (p. 31).

Minsky and Papert's 1972 report reveals the image of AI that was emerging in the MIT lab at that historical moment. The report supported the kind of nativism found in Piaget whose idea of "mental schema" was modelled on the Kantian "schemata." These schemata or mental schema were said to be the innate structures in children's minds that were used to process and make sense of perception. Papert and Minsky's Piaget-inspired constructionist (and ultimately Kantian) image of the infant mind substantially influenced computer science as it pursued symbolic AI for the next 50 years, rather than develop AI architectures predicated on neural nets. In fact, it was Minsky's and Papert's experimental psychological approach with infants and children that displaced the connectionist neural net models of AI developed by McCulloch, Pitts, Rosenblatt, and other AI researchers. It wasn't until the 21st century that neural net methods would return to dominate the field of computer science.

To be fair, Minsky and Papert embraced a grounded and embodied approach to learning, situating their study of children's reasoning in experimental design of simple tasks, creating learning environments where knowledge was said to be *constructed* rather than *instructed*. They argued that AI should look at infants and not experts when trying to design an artificial intelligence. They stated, "our main experimental subject worlds, namely the "blocks world" robotics environment and the children's story environment, are better suited to these studies than are the puzzle, game, and theorem-proving environments that became traditional in the early years of AI research" (p. 1). Our aim in this paper is not to critique Papert's constructionist pedagogy nor his aim to make mathematics and science education more imaginative and "concrete" with technologically rich environments. Instead, we hope to show how the philosophical foundations for his "constructionism" ultimately sprang from a Piagetian epistemology, where supportive learning environments were said to elicit children's "spontaneous" logic or geometry (Papert, 1991, p. 2). Borrowing from Piaget's experimental methodology, Minsky and Papert (1972) utilized a series of tasks designed to address and elicit specific cognitive skills in children, especially those skills related to visual perception and spatial sense, as these were considered skills that might be programmable in machines. Many of the experiments involved spatial analogies and attempts to discern separate objects within a scene, always starting with the prompt of asking the child to produce a "symbolic description" and thereby communicate and represent (a) their knowledge about the situation, (b) their process for problem solving, and (c) any methods of computation. The aim was to study these child-produced descriptions for how they revealed what Minsky and Papert called "heterarchical control structures." These structures were said to function as cognitive "problem-solving programs" that involved situated heuristics responsive to specific prompts. In effect, the heterarchical control structures were assumed to be symbolic and computational *a priori*, yet flexible programs innate to the infant mind and corresponding (*à la* Kant) to the material world. In other words, there was a presumed ontological correspondence between the built-in faculties of the human mind and the structural and physical workings of the world. Minsky and Papert understood the child as an agile agent with *a priori* programs. These built-in computational faculties were not necessarily consciously operationalized, but an "equivalent of such knowledge" was said to be obtained at the pre-reflective level:

> [s]uch an agent must know to a greater or lesser extent how to plan, produce, test, edit and adapt procedures. In short, it must know a lot about computational processes. We are not saying that an intelligent machine, or person, must have such knowledge available at the level of overt statements or consciousness, but we maintain that the equivalent of such knowledge must be represented in an effective way somewhere in the system.
>
> *(Minsky & Papert, 1972, p. 2)*

Analogical reasoning and "seeing" structure

Importantly, Minsky and Papert (1972) acknowledged that children's initial accounts were typically inexact and failed to offer formal rules. Consider, for instance, tasks that involve spatial analogical reasoning, and the request that the child select the appropriate choice to make the analogy in Figure 12.1. The task is to select one of the bottom row, according to the instruction: A is to B, as C is to which of 1, 2, 3, 4, 5?

The children's initial *justifications* for their choices were vague, and they failed to show how their analogical reasoning could become a rule that generalized to all similar cases. To address this inadequacy, Minsky and Papert asked the children for descriptions of what they saw, rather than for justifications of their choice. The emphasis on *description* – rather than justification – became crucial for Minsky and Papert because they believed that the process of generating a description would more effectively attend to the common structure that masks the tacit rule. In other words, they argued that a child's "mini-theory" can be articulated only if the child is first asked to "make up a description" and then asked to change that description so that it describes a new example (and thus the child generates an analogical rule for generalizing, without knowing they are

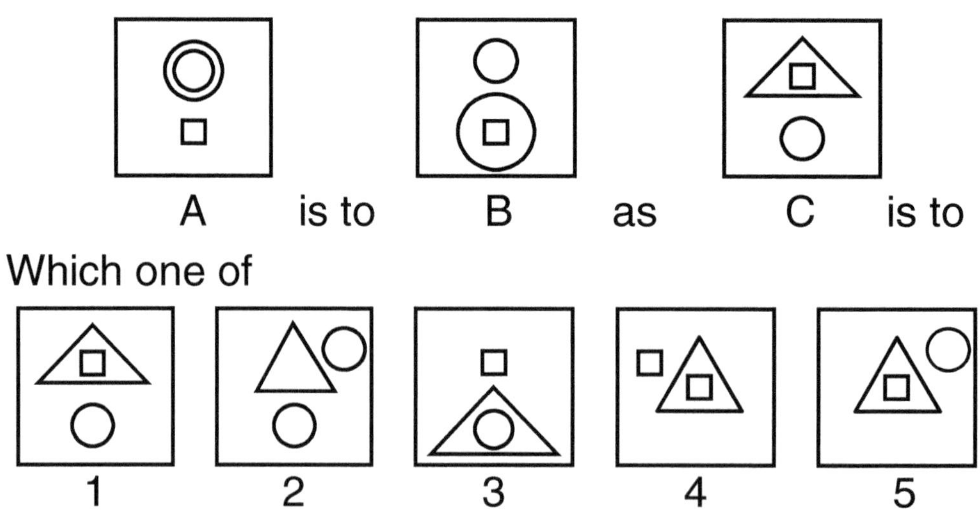

FIGURE 12.1 Analogical thinking (Minsky & Papert, 1972).

doing so). The experiments with children showed how analogical reasoning rested on the habit of formulating a verbal description of a state of affairs and transferring that description to a new state of affairs.

Minsky and Papert argued that machine intelligence should be crafted to mimic the infant mind. Rather than develop a connectionist approach that incrementally adapts through various kinds of network reasoning, Minsky and Papert sought a more qualitative image of learning, one in which learning occurred after only a few encounters, and where *the qualitative leap to structural learning* was characterized by innate *programs*. In other words, the use of language in the descriptions is simply the best medium to get to the more fundamental logical structure – that being the brain-based computable operation or program. For Minsky and Papert, programs are not merely strings of symbols but must include "primitive" symbols that represent some selection of the features and their relationship in a situation. Children came into the world with built-in primitive notions of space-time relationality – these were often represented as schematic flow networks. Notice that a description is always already a selection, a procedural (program-like) engagement. The authors claimed that "the description is itself a MODEL – not merely a name ..." (p. 4, caps in original).

To what extent do the ideas of Minsky and Papert provide an image of the infant mind that is neuro-symbolic? Despite their debt to Piaget and their trashing of the neural network methods, does their image of intelligence incorporate perceptual data? Notably, they assume that the children's thoughts are "an abstract data structure," which represents all kinds of features, relations, procedures, and other information. Children's verbal accounts are thus used here as an index of the symbolic (logical) process presumed to undergird analogical spatial reasoning. And yet, unlike the connectionists, this theory of learning downplayed the perceptual environment. Accordingly, Minsky and Papert dismissed the rival AI empiricist approach that emphasized neural network connectionist approaches, which was said to mimic human perception feed-back loops. Instead, they claimed that their research illustrated how children's descriptions *were not oriented* to perceptual cues, *and in fact were often at odds with these perceptual cues*. This was a crucial observation said to substantiate their theory of the infant mind. For instance, when asked to draw a cube, the children were said to ignore the perceptual cues and instead attend to important structural features, whereas the adults produced iconic representations of the cube ("iconic imitation") (Figure 12.2). Minsky and Papert thus argued that children override their senses and use highly schematic descriptions to represent geometric information. In other words, infant minds will consistently override perceptual cues and opt instead to "describe" the logical and causal structure. In the case of a cube, for

FIGURE 12.2 Child's drawing of a cube (left) versus adult "iconic" drawing of a cube (right) (Minsky & Papert, 1972).

instance, the adult draws for resemblance, while a child draws to capture key relationships (perpendicularity, number of faces).

It was these "formalist tendencies" in children's drawings that led Minsky and Papert to endorse a symbolic approach to AI, and not one that engages "more directly ... with the optical image" (p. 7). Although they acknowledged that children can be trained to draw and describe with "quantitative accuracy," they argued that "the symbolic mode is the more normal manner of performance" (p. 8). This finding was precisely what supported their belief in a symbolic AI, against the "empiricist" AI models that emphasized neural net incremental training on perceptual data.

The mind as program

Minsky and Papert's image of the infant mind as a "problem solver" was less hierarchical than previous Gestalt theories of learning where a presumed central processor would be said to manage data. Instead, they proposed a "heterarchy" where computational processes overlap or run in parallel. They state:

> we do not mean to suggest that our child had in his mind anything like the graphical image of his drawing, but rather that he has a structural network of properties, features, and relations of aspects of the cube, and that what he drew matches this structure better than does the adult's more iconic picture.
>
> *(p. 6)*

This heterarchical "structural network" is essentially a set of built-in cognitive programs that are flexible and responsive, radically at odds with a bottom-up stochastic model whose learning is based on large-scale sampling and searches. They were situating their work against empiricist projects like that of Willard Quine, and the wide-spread emphasis on behaviourism in the 1950s and 1960s. They saw their approach as an advancement from "abominable" behaviouristics models which assumed a blank-slate mind. We cast these behaviourists as "empiricists" rather than nativists, because they emphasized the empirical data in the learning process. In the contested space of the burgeoning AI field, Minsky and Papert argued that children had inborn symbolic faculties (programs) and that *description* was a way of helping them externalize – or make explicit – these programs.

Their ideas influenced learning theories for decades, fuelling the learning trajectory theories of the 1990s. Minsky (1988) will go on to characterize the mind as replete with "agents," further developing the mind-as-society theory with its emphasis on multi-agent modular brains, stating "minds are what brains do," while Papert (1980) will advocate for computer literacy and constructionist learning theories in education policy. Note that Papert's constructionist *pedagogy* was contrasted with "instructionism" and banking models of education (Papert, 1991); in other words, his theory of internal structural but flexible programs was not necessarily at odds with his pedagogy, which was focused on creating alternative learning environments where children might creatively develop STEM literacy through concrete and material activities (Papert, 1980; Turkle & Papert, 1990). His emphasis on "epistemological pluralism" was essentially a pedagogical and sociological position, exploring the distinctive ways that different contexts shape learning – this was an extremely important insight about learning, but his ontological assumptions were never adequately philosophically explained. In fact, the legacy of Kantian epistemology within this history of constructionist theories of learning has largely gone unexamined. Many working in this tradition are not aware of the historical provenance. Minsky and Papert's pedagogy rested on an epistemic-ontological assumption that there was a Kantian *correlation* between the presumed cognitive faculties of the human mind and the material conditions of the world. Such correlationism is traced back to Kant by Meillassoux (2008) (see Chapter 11 for more on this).

One can identify several tacit philosophical assumptions about minds and language in Papert and Minsky's approach. Firstly, Minsky and Papert assumed an unproblematic transparent translation between thought, language, and action, in their emphasis on descriptions. Despite their equivocation, that by "description" they did not mean "verbal description," but rather "an abstract data structure in which are represented features, relations, functions, references to processes, and other information" (p. 9), their experimental methodology ultimately rested on the child's *verbal and written* account of their thoughts. In other words, each experiment involved a request that the child explain what they were thinking, or describe their actions, or justify in words the choice they had made, or draw a diagram to capture the analogy, etc. Setting aside all the complex ways in which thought is not adequately captured by signification or conscious communication, Minsky and Papert went forth with a naïve realist approach to sign-making: "to find the secret, one has merely to ask any child to justify his choice" (p. 3). Secondly, knowledge was assumed to be *programs* that captured structural and logical relationships, conditioned by innate symbolic capacities rather than networked connectionist models. Minsky and Papert were adamant that their approach to knowledge and learning pointed to inherent cognitive faculties, which we have argued aligns with the Kantian tradition. As mentioned in the opening remarks and discussed in the next sections, this position is epistemically opposed to deep learning AI, where incremental deep training models are said to learn without built-in programs (Bengio, 2009).

Computational thinking

Minsky opened his ACM Turing Award Lecture in 1970 with the claim: "The trouble with computer science today is an obsessive concern with form instead of content" (p. 197). Fifty plus years later, we can make the case that Minsky's concern has come to dominate the subfield of computer science *education*. Minsky lamented the field's focus on formalisms, arguing that not enough was known about the basic phenomena of computing, and there was a need to generate universal theories of machine learning. Instead, he urged that a theory of computing would have to carefully attend to key foundational trade-offs (between memory size and computation time, serial versus parallel computing, addition and multiplication) and stay focused on the "computational content" of programs rather than their syntactic conditions. Most importantly, he discussed examples of children learning mathematics to support his claim that programs and procedures correspond to human learning skills. Computing, states Minsky, is itself the best model for the mind:

> We cannot digress here to answer skeptics who feel it too simpleminded (if not impious, or obscene) to compare minds with programs ... to model his own self [the child]

> cannot use the engine or the organism or the city or the telephone switchboard; nothing will serve at all but the computer with its programs and their bugs.
>
> *(Minsky, 1970, p. 205)*

Indeed, he is confident that cognitive scientists will eventually show how children problem-solve by enacting common computing procedures, and lists examples of these procedures: calling sequence, dispatch table, error message, compiler, hash coding, decision tree, diagnostics, and asynchronous processor. He claims that "before computation, the community of ideas about the nature of thought was too feeble to support an effective theory of learning and development" (p. 208).

Minsky's concern was that computer scientists and, especially, computer science educators were caught up in technical formalisms and neglected the more fundamental questions about computing and cognition. The concept of "CT" emerged as a response to Minsky's critique. Since the 2000s the charge to teach "CT" has been the rallying cry to include "coding," computer programming, and computer science more generally in every level of education, with a special focus on introducing CT as early as possible in K-12 education. This renaissance of computer science education in the guise of CT stems from Jeanette Wing's 2006 article for the *Communications of the ACM* (Wing, 2006), where she poetically rushes across the range of computer science disciplines and recasts them as essential, general tools for critical thinking and problem solving. Wing (2008) identifies *abstraction* as the key concept of CT and urges for the inclusion of CT across disciplines and *for everyone* because computing touches all of our lives in myriad ways. As Wing's CT began to take shape in curricula and instructional methods, Grover and Pea (2013) summarized the key aspects of CT they identified in the literature. They found the following nine principles as common and consistent to conceptions of CT (Grover & Pea, 2013, pp. 39–40):

1. Abstractions and pattern generalizations (including models and simulations)
2. Systematic processing of information
3. Symbol systems and representations
4. Algorithmic notions of flow of control
5. Structured problem decomposition (modularizing)
6. Iterative, recursive, and parallel thinking
7. Conditional logic
8. Efficiency and performance constraints
9. Debugging and systematic error detection

These nine principles are meant to reflect the interdisciplinary nature of computer science, drawing on expert reasoning in mathematics, engineering, scientific process, information theory, and design. Subsequent literature on CT largely stays within the bounds identified by Grover and Pea (see Denning & Tedre 2021 for a summary). There are indeed continuities between the symbolic image of the infant mind and this image of CT, but there are also new emphases which reflect a growing interest in more incremental data-hungry images of AI. Abstraction was said to be the "keystone" of CT and as such distinguished computing from other kinds of thinking (Grover & Pea, 2013, p. 39). However, the CT concept of abstraction was quite different from that of Papert and Minsky, as it pertained to *pattern recognition* rather than *analogical reasoning*. The primacy of abstraction in CT reflects a move away from computer programming and algorithms as the central components of computer science, away from abstraction as structural and compositional, and towards an abstraction that is subtractive and in the service of neural network pattern recognition.

Prior to CT, abstraction in computer science was closely aligned to Piaget's schema and Papert's powerful ideas – the programmer attempted to express universal concepts in code. The influential *Design Patterns: Elements of Reusable Object-Oriented Software* (Gamma et al., 1994) elaborates a "pattern language" to express these universals and offers an initial collection of abstract patterns. In the introduction to his popular college programming textbook, *Thinking in Java*, Bruce Eckels argues, "the complexity of the problems you're able to solve is directly related to the kind and quality of abstraction" (2006, p. 15). Eckels' abstractions reflected underlying essential qualities that the programmer strove to perceive. CT offers a new image of abstraction (as pattern recognition rather than analogical reasoning) where abstractions exist "out there," as though they were emergent waves in a sea of data. Abstraction emerges as a pattern in statistical distributions, rather than the kind of abstractions said to occur when children discern structural relationships. This is also a methodological shift away from studies of children's behaviour and towards studies of expert behaviour. Ironically, the shift to "CT" represents a shift away from Minsky and Papert's developmental models of child cognition and towards the study of *expert* reasoning practices. When Wing wrote, "Thinking like a computer scientist means more than being able to program a computer. It requires thinking at multiple levels of abstraction," she is using expert practices as the model of CT.

Pedagogy and capital

It's difficult to argue that Wing's vision has not garnered success. Bibliometric analysis (Figure 12.3) of the prevalence of "CT" in the Scopus databases of scholarly literature indicates zero occurrences of CT prior to 2006 and then slow growth in the literature, after Wing's seminal piece, from 2006 to 2011. The SCOTUS search only looked at titles, abstracts, and keywords in scholarly documents, not the full text of those documents.

The year 2011 appears to be a turning point where we see a large jump in the use of the term CT: published documents grow from 58 in 2012 to 794 in 2022. Much of this growth should be attributed to the adoption of CT by professional organizations like the International Society for Technology Education (ISTE) and the Computer Science Teachers Association (CSTA). More than any other group, though, the non-profit Code.org (founded in 2012) became the well-funded lobbying arm for the paradigm of CT. Code.org mobilized its nearly $60M in capital, largely donations from big tech and private equity, to spread the word of CT. Together these groups seized the idea and popularized it through the dissemination of educational materials, grants for schools and teachers, formalized teaching standards, political lobbying (for State education standards and certifications), and standardized exams like the new Computer Science Principles Advanced Placement (AP) exam from the College Board. Although this quick analysis of Scopus shows zero references, we know that Wing did not coin the phrase. Papert may have been the first to use the term in *Mindstorms* (1980, p. 182), but he writes "computational culture" and other constructions much more frequently to describe his work. He only uses CT once. Turkle and Papert (1990) return to CT to briefly discuss the limits and problems of CT, but mostly

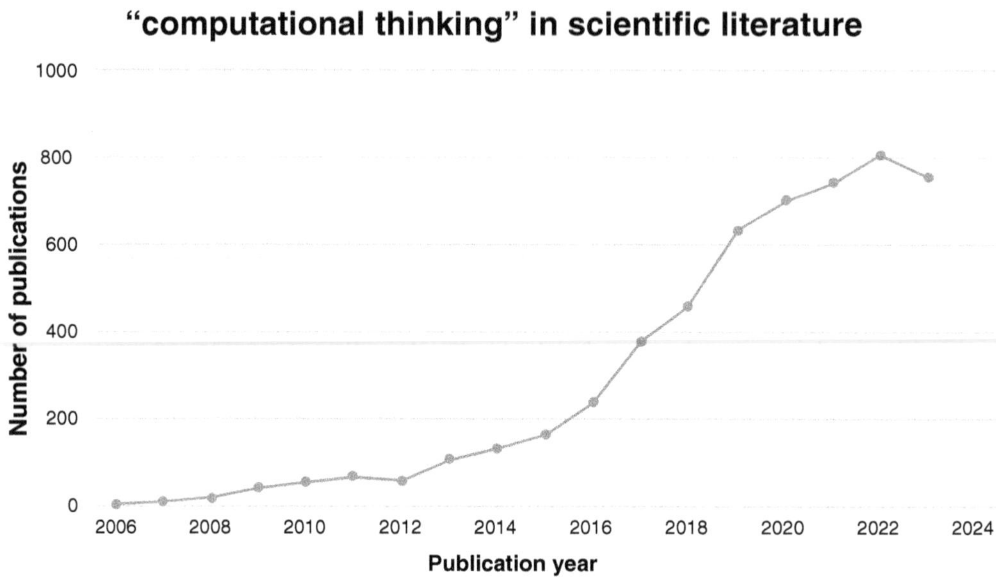

FIGURE 12.3 "Title, abstract, keyword" search metrics for "computational thinking" in Scopus.[1]

in terms of classical images of logical/formal reasoning, and not in terms of the above list of nine "practices" (Turkle & Papert, 1990). Moreover, their epistemic pluralism reflects a kind of socio-cultural relativism focused on diversity of programming style and expression, and does not engage more radically with a post-Kantian or posthuman philosophy of mind and machine associated with generative neural networks and discussed below (Braidotti & Fuller, 2019).

We can see how the widespread success of CT radically shifted the epistemological commitments within the field of the Learning Sciences, and also *tacitly* reflects a shift away from more classical approaches to AI. To what extent do we see here a new image of thought? In what way has the symbolic image of the infant mind as a learner been displaced by an image of a network mind? Wing claims that CT underpins the intellectual processes needed for all fields of study today, and offers a way of thinking and approaching problems for everyone from STEM to humanities students. Accordingly, proponents argue CT is the dominant, necessary thinking for our current age because it *powers* technological change and *anticipates* societal needs. However, Wing's examples hew closely to computing and STEM fields: statistics, physics, astronomy, computational biology, and economics (Wing, 2008). In other words, CT has been leveraged to support state-sanctioned science and technocratic social orders. We find that CT's content-free formalism and operationalized problem-solving make it a close ally for the dominant powers of late-stage capitalism.

Prior to Wing, Papert was reluctant to cast the power of computing in terms that were overly shackled to a utilitarian problem-solving approach. While he would have agreed that computational approaches are closely related to both mathematics and scientific experimentation, their true value was to the fields of psychology and philosophy. According to Papert, computing offers us, collectively and individually, a window into how we *think* and *know*. The study of computing and learning for Papert and his team, under the guise of cognitive science, was an epistemological project. Specifically, Papert (1980) was concerned with an epistemology of "powerful ideas" – foundational ideas that help us organize our understanding of the world. These ideas were logical, mathematical, geometric, reflecting core knowledges which could be better understood through the lens of computing. Papert's epistemology represents a strong contrast to Wing's operational CT. With the Piaget-inspired constructionist approach to machine learning, AI research was meant to involve a deep study of conceptual networks or structures, reflecting core logical and geometric relations. While Wing saw an image of human flourishing in the principles of CT, Papert claimed that computing offered an alienation of the self that allowed us to probe, test, and refine the limits of our own knowledge. This part of his project, however, is rarely, if at all, acknowledged, as it complicated the Constructionist political project, which aimed (whether effectively or not) to get computers into schools and redistribute resources to disenfranchised communities. Papert hoped that computers would undermine authoritarian models of instruction and unlock organic, emergent learning in children, no matter their context. Where Wing's CT seeks an accretion of power to computing and STEM, Papert envisions a Trojan Horse (Papert, 1997) where computing enables the bricolage, hard fun, and aesthetic experience that create the foundation for an emergent, self-realized learning. And yet both CT and constructionism seem poorly equipped to reckon with our current computational culture, where connectionism and vast interleaving neural network algorithms rule supreme.

New software empiricisms

Returning to the epistemological debates between nativism and empiricism, we conclude in this section by suggesting that Software Studies offers a kind of post-Kantian new empiricist position well-suited to the current emphasis on deep learning AI. In other words, we note that as AI research has shifted away from symbolic nativist images of the infant mind towards more explicit neural network approaches to cognition, software studies has emerged and with it a new kind of empiricist epistemology. Fuller and Weizman (2021) deploy an "investigative aesthetics" to exert an "aesthetic power" made possible by the vast amounts of crowd-sourced sensory data "sensed" today. They name this approach *Forensic Architecture* and apply the approach in human rights investigations to uncover abuses, such as the use of tear gas against civilians. Their technique uses advanced machine learning, neural networks, and computer vision to scour vast amounts of images and video uploaded to YouTube, Twitter, and other social media. Their software helps build timelines and evidence of human rights abuses, leading to a new form of creative work like Triple Chaser (Forensic Architecture, & Poitras, 2019), which is equal parts documentary film, criminal prosecution, and artificial intelligence lecture. This work exploits the hardware, software, networks, and methods of big tech and State surveillance to advocate for justice for marginalized groups. This involves a "rescue" or harvesting of data traces buried "beneath layers of erasure" (Fuller & Weizman, 2021, p. 57).

Their approach disrupts Enlightenment Kantian assumptions about aesthetics by studying sensory experience as an exo-somatic aspect of space-time, rather than that which is synthesized within the human mind. Indeed, Fuller and Weizman (2021) redefine computational aesthetics as a post-Kantian aesthetics. Rather than assume cognitive faculties which are the conditions of sensory experience, their image of thought rests on an expanded notion of aesthetics which "does not seek to silence human sensibility but rather augment, play with and test it, to expand and open the conversation to other forms of sensing and sense-making" (p. 44). This posthuman approach to expanded sensibility demands a different empiricist epistemology that allows them to engage with technical being or technical machine intelligence beyond the mirror of the infant mind. They present a series of projects that involve methods of "hyper-aesthetics" explaining that "to hyper-aestheticize is to heighten, elicit or exacerbate the capacity of bodies, technologies or states of matter to sense and increase perceptual experience; it entails an increase in sensitivity and can perhaps augment a capacity to care" (Fuller & Weizman, 2021, p. 58). This approach offers a more-than-human ecological image of thought, quite distant from the individuated nativist image of the human mind processing sensory input. Similarly, Braidotti and Fuller (2019) suggest that generative machine learning algorithms be considered *posthuman* precisely because they engage in "processes of abstraction that may in turn generate grounds of operation that are outside of the original conditions abstracted from, and thus producing novelty" (p. 9). Here, abstraction lives in the network but has gone beyond pattern recognition, evoking some of the more *compositional* aspects of abstraction studied by Papert and Minsky (Buckner, 2019).

This new practice distributes thought across an information connectionist network and turns away from prior onto-epistemic assumptions of innate cognitive structure. Insofar as cognition becomes exterior or exosomatic or ecological, it has taken the form of a "network," which has replaced all previous images of knowledge, learning, and computing. For Galloway (2021), this follows directly from what he calls the "behaviourism" of the last century and its relinquishing of interiority. The interiority of the learning child was crucial for the early Piaget-inspired symbolic image of machine intelligence. The very concept of the interior brain "program" as a dominant trope for intelligence, as articulated by Minsky and Papert, is displaced by a socio-material network that is no less opaque for being external to the individuated mind. Galloway's concern, like many others, is with the lack of legibility of this mode of network intelligence, and with the black box processing and its apparently inexplainable emergent learning. Minsky and Papert's Kantian children have become part of a massive transindividual network that learns through obscene amounts of data. Obfuscation moves from within the interiority of the mind, where phantom Piagetian "schema" were supposed to exist, to the vast undulating hypothesis space, traversed by the blind neural net algorithms and their method of gradient descent.

What does this mean for CT? Where do Papert's powerful ideas live now? If banished from the murky interiority of Kantian cognitive models, are these orphan ideas now circulating in the network somehow? Are they on the outside, spread so thin we can barely see them, or emergent across an affinity network, branded and monetized? Powerful ideas need fluid mutable media if they are to remain powerful. Is cognition hiding, lurking in the shadows of our software networks, flourishing in the dark? Is there a shadow network, a kind of "thought network" that rises to Simondon's (2017) more hopeful rendering of a "technical mentality" where human and machine partake in a mutual reciprocity?

> Here, the technical mentality successfully completes itself and rejoins nature by turning itself into a thought-network, into the material and conceptual synthesis of particularity and concentration, individuality and collectivity—because the entire force of the

> network is available in each of its points, and its mazes are woven together with those of the world, in the concrete and the particular.
>
> *(Simondon, 2012 p. 9)*

For Galloway (2021), this new image of the external network-mind is directly linked to software because software is *inherently* opaque and operates in a "network form." Software "acts like a network, a network of message-sending" (p. 228). This opacity is built into the ouroboros-like complexity of software, and its transitory nature, as it continuously switches back and forth between the source code, the executable application, and the runtime interface. The source code is a programming language, a human-readable text, while the executable application runs on machine-readable code compiled from the source, and the runtime interface is "the existing experience of software, as it relates to a user" (p. 227). Software slips in and out between these three aspects, according to a black box logic; in other words, software architecture shuttles semantics here and there across these layers, hiding meaning behind data and pattern: "software hides itself at exactly the moment when it expresses itself most fully" (p. 228). It is this intrinsic obfuscation that has led to a widespread crisis in legibility. To address this issue of legibility, he argues that Software Studies should pursue methods like historical genealogy and reenactment to expose and make material what we might call "the computing content" of the algorithmic architecture.

This approach to Software Studies shares with Fuller and Weizman a commitment to disrupt software obfuscation and opacity, attending to questions of the legibility and explainability of AI: "The politics of the new millennium will thus be a politics not so much of time or space, but of appearance" (Galloway, 2021, p. 235). Galloway mentions Hito Steyerl's extremely influential installation video "How not to be seen," Simone Browne's *Black Luminosity*, the French anonymous anarchist group *Invisible Committee*, and various 21st century movements that leverage new kinds of in/visible exploit. Such exploits perform a kind of subtractive politics, a politics of opacity and refusal. He shows how computation emerges in different ways depending on the media, and how "the digital" thrives and atrophies over time, suggesting that we must do both critique and computation, both critical analyses and "algorithmic reenactments" if we hope to unfund some of the more harmful CT in our midst. He discusses a series of case studies in the history of computation that shed light on our current algorithmic condition, selecting a set of somewhat unusual and provocative stories that read like character studies or portraits of radical computational lives (Barricelli's cellular creatures, Debord's board game, Dietz' algorithmic textiles, etc.). The case studies offer up examples of alternative experimental practices that emerged alongside the shift to distributed network knowledge in "the long digital age."

We will have to keep this kind of politics afloat when the frustrations with legibility become too much, and we tire of our deep learning models and their free-wheeling dance with data and "cargo cult AI" (Levine, 2023). At that point, the obfuscation of software may become too much, and the data-hungry neural network AI models may be cast out, or at least bracketed, as we seek once again the Kantian programs of interior judgement. Perhaps then the neuro-symbolic infant mind will come charging back, performing intuitive feats, and displace the exosomatic image of eco-cognition sustained in Software Studies. Perhaps our network image of mind will dissipate, and once again we will be asked to accept the opaque interior workings of program-like cognitive schema. This might well happen sooner than later, especially as the limitations of deep learning AI become increasingly obvious (Shanahan, 2024). Indeed, we hear Papert-like claims that the conceptual and material world is co-composed of "programs" that correlate (Ellis et al., 2020). And we note the continued influence of cognitive psychologist Elizabeth Spelke (2017) on AI theorists, as she argues that there are innate "core knowledge" structures that determine learning progressions. Spelke argues that there is a set of six cognitive domains of which infants seem to have intuitive (i.e. innate) knowledge. These are (1) Objectness, (2) Number (counting), (3) Geometry and space, (4) Agentness, (5) Place, and (6) Social relations. The debates between the nativists and the empiricists continue, and although Spelke's specific cognitive theories diverge from those of Piaget, we should nonetheless place her on the nativist side, as she diverges more significantly with socio-political turns in the Learning Sciences. And it is this nativist perspective that continues to stand as the foundation of most psychometric benchmarks for spatial and logical reasoning in children and machines. Attempts to build machine intelligence as neuro-symbolic draw explicitly from this work, including the 2019–2025 proposal ARC-AGI (Abstraction and Reasoning Corpus), which cites Spelke's cognitive framework of "core knowledge" as a basis for shaping future machine intelligence (Chollet, 2019). Such benchmarks and other related tests show the poor learning skills of neural network models – including those using transformer algorithms – compared to the agile and rapid learning skills found in children. Despite the success of various "empiricist" deep learning algorithms, further enhanced with generative strategies, the nativist theories of the infant mind persist in AI debates today.

Note

1 Scopus is a database of scholarly works that contains titles, abstracts, and keywords of publications. This graph presents data from a Scopus search on December 4, 2023. Accordingly, not all citations for 2023 are counted. The Scopus database is not exhaustive but *is* representative of the terms searched.

References

Bengio, Y. (2009). Learning deep architectures for AI. *Foundations and Trends in Machine Learning*, 2(1), 1–127.

Braidotti, R., & Fuller, M. (2019). The posthumanities in an era of unexpected consequences. *Theory, Culture & Society*, 36(6), 3–29.

Buckner, C. (2019). Deep learning: A philosophical introduction. *Philosophy Compass*, 14(50). https://doi.org/10.1111/phc3.12625

Chollet, F. (2019). *On the measure of intelligence*. https://doi.org/10.48550/ARXIV.1911.01547

Denning, P., & Tedre, M. (2021). Computational thinking: A disciplinary perspective. *Informatics in Education – An International Journal*, 20(3), 361–390.

Eckels, B. (2006). *Thinking in Java*, 4th edition. Hoboken, NJ: Prentice Hall.

Ellis, K., Wong, C., Nye, M., Sable-Meyer, M., Cary, L., Morales, L., Hewitt, L., Solar-Lezama, A., & Tenenbaum, J. B. (2020). *DreamCoder: Growing generalizable, interpretable knowledge with wake-sleep Bayesian program learning*. https://doi.org/10.48550/ARXIV.2006.08381

Forensic Architecture, & Poitras, L. (Directors). (2019). *Triple chaser* [Documentary, Short]. London: Praxis Films.

Fuller, M., & Weizman, E. (2021). *Investigative aesthetics: Conflicts and commons in the politics of truth*. New York: Verso.
Galloway, A. R. (2021). *Uncomputable: Play and politics in the long digital age*. New York: Verso.
Gamma, E., Helm, R., Johnson, R., & Vlissides, J. M. (1994). *Design patterns: Elements of reusable object-oriented software*, 1st edition. Boston, MA: Addison-Wesley Professional.
Grover, S., & Pea, R. (2013). Computational thinking in K–12 a review of the state of the field. *Educational Researcher*, 42(1), 38–43.
Lake, B. M., Salakhutdinov, R., & Tenenbaum, J. B. (2015). Human-level concept learning through probabilistic program induction. *Science*, 350(6266), 1332–1338.
Levine, E. V. (2023). Cargo cult AI. *Communications of the ACM*, 66(9), 46–51.
Meillassoux, Q. (2008). *After finitude: An essay on the necessity of contingency,* Pbk. edition. New York: Continuum.
Minsky, M. (1970). Form and content in computer science (1970 ACM turing lecture). *Journal of the ACM*, 17(2), 197–215.
Minsky, M. L. (1988). *The society of mind*. New York: Simon and Schuster.
Minsky, M., & Papert, S. A. (1972). *Artificial intelligence progress report*. Available at https://dspace.mit.edu/handle/1721.1/6087
Papert, S. (1980). *Mindstorms: Children, computers, and powerful ideas*. New York: Basic Books, Inc.
Papert, S. (1991). Situating constructionism. In I. Harel & S. Papert (Eds.), *Constructionism* (pp. 1–10). Westport, CT: Praeger.
Papert, S. (1997). Why school reform is impossible. *The Journal of the Learning Sciences*, 6(4), 417–427.
Piaget, J. (1954). *The construction of reality in the child.* (Trans. M. Cook). New York: Basic Books.
Shanahan, M. (2024). Talking about large language models. *Communications of the ACM*. 67(2), 68–79.
Simondon, G. (2012). Technical Mentality (Trans. A. De Boever). In A. De Boever, A. Murray, J. Roffe, A. Woodward (Eds), *Gilbert Simondon: Being and Technology* (pp. 1–15). Edinburgh: Edinburgh University Press.
Simondon, G. (2017). *On the mode of existence of technical objects*. (C. Malaspina and J. Rogove, Trans). Minneapolis, MN: Univocal. (Original work published 1958).
Spelke, E. S. (2017). Core knowledge, language, and number. *Language Learning and Development*, 13(2), 147–170.
Spelke, E. S. (2020). What infants know (and don't know). CVPR workshops on minds vs. machines: How far are we from the mind of a toddler? Seattle, June 15. Talk available at https://www.youtube.com/watch?v=r-F8Dlp-POk
Turkle, S., & Papert, S. (1990). Epistemological pluralism: Styles and voices within the computer culture. *Signs*, 16(1), 128–157.
Wing, J. M. (2006). Computational thinking. *Communications of the ACM*, 49(3), 33–35.
Wing, J. M. (2008). Computational thinking and thinking about computing. *Philosophical Transactions of the Royal Society A: Mathematical, Physical and Engineering Sciences*, 366(1881), 3717–3725.

13

THE SPATIAL LOGIC OF EPISTEMIC IMAGINARIES

Guided by Édouard Glissant

Introduction

The term "epistemic imaginary" underscores the power of the imagination in sparking insight, shaping knowledge practices, and establishing epistemic authority. The imagination is the milieu within which reason habituates itself, fabricating hypotheses and anticipating empirical encounters. This complex polemic between the imagination and empirical "correction" gradually sustains habits of trust and epistemic certainty. An epistemic imaginary is thus a set of knowledge claims performed passionately as a truth warranting, a set of declarations that establishes positionality and situates perspective, perhaps substantiated by other thinkers in some degree, and often enshrined and then imposed on others beyond the initial pertinent boundaries. An epistemic imaginary is always made manifest in a manifold of contested assertions, delusions, and speculative ventures, almost always situated in spatial politics. *Knowledge and truth-telling are habits born of the imagination in place.*

Critical theories of various kinds start in the contested authority of epistemic imaginaries, where they critique the current habits of truth-telling and generalization. To be critical is to first reckon with (resist, reform, relay) the current practices and historical aftermath of claims to epistemic authority, which is why critical theory has traditionally occupied the outsider position, standing up for a truth that has not yet been encoded within current regimes. To stand up and speak against the violence of an abusive epistemic authority is essential; a critical perspective exposes current cultural configurations inscribed into the knowledge practices and empirical habits of past generations, carried forth without consideration for how these come at the cost of lost life and destroyed habitat. Such critical approaches set the stage for change, priming the conditions, but they remain empty gestures if not also forged with creative impulse, sparked into action and risk-taking by the speculative power of the imagination.

It is the power of the imagination (speculation, fabulation, creative alteration, etc.) that ensures criticality can engender the new. The worn out and spent critique of skeptics who will not cast their lot in the name of any creative conjecture, for shame of failure, serves

DOI: 10.4324/9781003538769-13

in the end to support the culture-mulching machines which they aimed to overthrow. The challenge is to sustain the critical work while ad/venturing into the turbulent unknown of creation, to wade out into the metamorphic earth re-imagined, into a space of potentiality that engenders as much as it corrects.

This chapter works with the ideas of the French Caribbean poet and place-philosopher Édouard Glissant (1928–2011) as a guide in the murky poetic realm of epistemic imaginaries.[1] I will explore the critical-creative power of current epistemic imaginaries, moving from the necropolitics of Achilles Mbembe (2019) to the speculative futures of Kara Keeling (2019), both of whom respond to the powerful writings of Glissant in different ways, as they try to reshape the space of "earth dwelling" and epistemology in a postcolonial world. For each of these thinkers, the dangerous power of speculation is enlisted, not only to critique the over-ambition of epistemic authority, but to open up new imaginaries and alternative spatial logics.

This chapter also thinks with the ideas of Gilles Deleuze (1925–1995), who plays a significant role in the writing of Glissant, Keeling, and Mbembe. Both Keeling and Mbembe work closely with Deleuze's texts, relaying his work on capital and control societies into postcolonial contexts. Moreover, Deleuze and Glissant are linked by their shared commitment to rhizomatics, nomadicism, creolization, and other "baroque derangements" of late capitalism. According to both Deleuze and Glissant, thought is a thoroughly spaced medium, and spaces itself out into the world, so that knowledge practices are always linked to spatial logics and imaginaries. In particular, Glissant (1997, 1999, 2019) asks that we break up the controlling continental sorting of center-periphery spatial logic, aiming instead to scatter life across an archipelagic network of links and relays, a rhizomatic lateral connectivity that risks the offer of a temporary archipelagic arrangement. In this way, Glissant offers an alternate earthly belonging that is characterized by the archipelagic rather than the continental and its colonial empires; he pursues an onto-epistemology of relationality that inherits the violent legacy of colonialism. He dives deep into the folds of "thinking thought" and puts himself at risk, imagining a new image of earthly thought:

> Thinking thought usually amounts to withdrawing into a dimensionless place in which the idea of thought alone persists. But thought in reality spaces itself out into the world. It informs the imaginary of peoples, their varied poetics, which it then transforms, meaning, in them its risk becomes realized.
>
> *(Glissant, 1997, p. 1)*

What kind of epistemology can adequately honor the profoundly unknown composition of multiple earths, shaped by an infinity of heterogeneous agencies and waves of action that defy globalism? How might we manage the pluralism of collective truth-telling practices that proliferate, each marshaling different modal degrees of authority, an authority expressed through geolocation and economic-colonial land grabs, as well as experimentation, speculation, proof, and poetic method? These diverse methods of truth-shaping tap the power of the imagination, for good and bad. Epistemic imaginaries adequate to the contingencies of the Anthropocene, Anthrobscene, Capitalocene, and Plantationocene will have to reckon within newly altered dependency relations where trust must be re-established, and where the collective credibility of truth claims must be reconsidered under posthuman conditions. If an epistemic imaginary is always in relation to "convoked authorities" with associated collective beliefs, achieved by way of rituals, attention, care, violence, and in/ex/clusion, how must we act in order to understand?

Credibility and part-whole relationality

Current epistemic imaginaries dovetailed with immanence, agency, and animacy must engage in a risky terrestrial diplomacy under "post-truth" conditions (de Freitas & Truman, 2020). The current climatic regime reveals to those who were previously unable to see how the metamorphic earth rightfully refuses epistemic imaginaries that are situated nowhere, detached and exceptional in their desire for control. If past empiricisms have nonetheless been our most powerful epistemic engagement in certain practical matters, as well as platforms for mobilizing violent war machines, they have achieved that status through complex assumptions about human and more-than-human agency (Jackson, 2020). If a new more situated empiricism is possible (Stengers, 2017), it will have to carry on increasing the visibility of non-human agency (volcanoes, bacteria, electrons, etc.), enhancing the diversity of a posthuman politics, while re-establishing the credibility of these new accounts.

By multiplying agencies, onto-epistemic practices produce the real, rather than "socially construct" the real; this distinction is crucial, since to "socially construct" fails to reckon with the material and generative force of that which is not social. Human exceptionalism struggles to imagine an outside to the social, and social constructivist projects have tended to read all problems in terms of signification (linguistics, semiotics, code), dwelling on the existence of a veil or mask that separates humans from the real. Scholars in the history of science, technology, and society (STS) have argued that such "critique has run out of steam" (Latour, 2004, p. 225) if all it can do is show how past human practices were trapped in language, and wrongfully justified by reference to truth warrants that were utterly fabricated (Latour, 2004). To be clear, such critique is necessary, but not sufficient. There is a danger of flattening the landscape into alternatively "socially" constructed truths, a terrain lacking any collective means for discerning credibility or reckoning with the powers of the speculative.

As Deleuze and Guattari (1994) intone, "Perspectivism, or scientific relativism, is never relative to a subject: it constitutes not a relativity of truth but, on the contrary, a truth of the relative..." (p. 130). This point bears repeating: a mere relativity of the truth fails to encounter (or imagine) difference in itself – because it can only achieve an affirmation of alternative truths, without gathering them together. There is a stronger affirmation of difference in the concept "difference in itself," which is precisely how Deleuze and Guattari argue for the ontological weight of "the truth of the relative" and not merely the relativity of truth. There needs to be a way of investigating the truth of the relative as "a property of the world itself and not only a phenomenon of language about the world" (Latour, 2017, p. 69). Indeed, the terms "situated" and "context" often hinder our ability to speak to the truth of onto-epistemic inseparability and entanglement. To be entangled is more than simply recognizing the relativism of truth claims, and more than merely a reference to the situated nature of knowledge; entanglement is about the ontology of relationality. It is this quasi "realist" ontology of dependence relations that a theory of knowledge (at its best) pursues and elaborates.

Édouard Glissant does this kind of work, seeking a kind of poetic empiricism in the afterlife of slavery, exploring the links between the imagination and knowledge. He seeks a postcolonial knowledge of relationality, sensitive to what he calls the "tout-monde," which is a new way of thinking "whole" that is not an imperial totality:

> Not just a specific knowledge, appetite, suffering and delight of one particular people, not only that, but knowledge of the whole, greater from having been at the abyss and freeing knowledge of Relation within the Whole.
>
> *(Glissant, 1997, p. 8)*

This "Relation within the Whole" is not a mixture or a melting pot. He also uses the term "chaos-monde" to refer to our living with contingency, incompleteness, and uncertainty:

> The aesthetics of the chaos-monde (what we were thus calling the aesthetics of the universe but cleared of a priori values) embraces all the elements and forms of expression of this totality within us; it is totality's act and its fluidity, totality's reflection and agent in motion. The baroque is the not-established outcome of this motion. Relation is that which simultaneously realizes and expresses this motion. It is the chaos-monde relating (to itself). The poetics of Relation (which is, therefore, part of the aesthetics of the chaos-monde) senses, assumes, opens, gathers.
>
> *(Glissant, 1997, p. 94)*

Glissant offers us a new way of conceptualizing part-whole relationality, a new way of belonging to the earth. He is concerned that the violence of filiation is all too often buried in the sacred mystery of the root, a mystifying and mythologizing process that operates through a negative kind of mystification of truth. Such mythologizing of root and origin often functions in nationalistic regimes, which demand a reductive transparency around identity:

> Against this reductive transparency, a force of opacity is at work. No longer the opacity that enveloped and reactivated the mystery of filiation but another, considerate of all the threatened and delicious things joining one another (without conjoining, that is, without merging) in the expanse of Relation.
>
> *(Glissant, 1997, p. 62)*

Glissant affirms an "earth aesthetics" in order to pursue a new kind of knowing better suited to postcolonial spatial logics. But he also demands new more helpful words – we might stray from the awkward term "people," for instance, as "we" attempt to "belong" to land and language without appeal to root. In search of such alternatives, he speaks of the "crowd" as both rain cloud and peaceful smoke, a gathering that dwells in poetic murmur (Glissant, 1997, p. 9). He then turns to the practice of knowing through errantry, a mode of gathering that is not the heavy-handed "comprendre" in which learning is a kind of "grasping." He asks: How might we gather together and sustain this sense of dispersal, not as "ground" but as "errantry," a wayward movement that stretches well beyond the negative positioning of "exile"? How might a collective gathering of nomadicism be sustained, which escapes the arrow and circular modes of nomadic subjection and nomadic imperialism? How to begin in the archipelago, in a diffractive Caribbean Sea, when there has been so much forced movement and enslavement before finding oneself "here"? How to carefully recall the history of those torn from their homes and deported to this place (any place), in the enslaved service of global capitalist trade routes and knowledge-seeking moderns? In the next sections, I show how Achilles Mbembe and Kara Keeling use these insights from Glissant to address what they characterize as a current crisis in epistemic practice and "reason."

Necropolitics and new terrestrial conditions

Necropolitics (Mbembe, 2019) began as a lecture and came to be published as a powerful critique of Foucault's biopolitical reading of power and a simultaneous affirmation of Glissant's poetics of Relation. Mbembe (2019) explodes the spatial politics of "our times," as seen from the location of South Africa, announcing his "transversal" critical methods as "opening, crossing, and circulation" (p. 10). Concerned with our "exit from democracy" and apparent departure from "reason," and the current "ordeals" of the world, Mbembe describes four contributing factors: The first factor is global migration, a "demographic transition now under way" characterizing the "modern" period, due to the forced resettlement of peoples from the south, as part of the long history of colonialism, slave trading, and the predation/extraction and redistribution of wealth. This redistribution of people and wealth entails both a "trading-post colonialism and settler colonialism," whereby the differences between these two colonialisms, in terms of specific impact, pertain to the role of the state and the new "swarming," which adds to previous waves of migration (p. 13).

The second factor for Mbembe is the transition "from the human condition to the terrestrial condition" (p. 13). This new terrestrial condition harkens a post-digital "general ecology" and a continuance of geometrics through more than human means, producing a "broadened geography, one that is spherical and irreversibly planetary" (p. 13). This also involves then a plastic human, and a linked notion of the digital subject. The third factor concerns this intrusion of computational machines into all aspects of life, fragmenting identity into a delicate crystalline subject. Citing the geographer Claire Larsonnaire, he states:

> It's no longer through the face-to-face encounter with the portrait, or the figure of the mirroring-presenting double, that the subject is tested, but through the construction of a form of presence of the subject closer to tracing and projected shadow.
>
> *(p. 14)*

Finally, the fourth factor is the configuration of capital in the financial markets, and their use of artificial intelligence (AI) technology, allowing finance to become spectacularly fragile and yet capable of marshaling unlimited power of destruction. According to Mbembe, it is these four elements together that threatens democracy, a democracy that is always burdened by its historical foundation on pro-slavery capital, which he names "the nocturnal body of democracy."

This current transitional economic-ecology harbours within it the ongoing destructive power of past violence: "Democracy, the plantation, and the colonial empire are objectively all part of the same historical matrix" (p. 23). But the historical trace of this colonial conquest has become part of the fabric of a new kind of "unregulated war" or "war outside the law" whereby police states and control societies work to mask the underlying inequity and injustice (p. 25). Colonialism "bears democracy, often in the guise of a mask" (p. 27). This originary violence that undergirds our current condition is typically exported "to third places, to nonplaces, of which the plantation, the colony, or, today, the camp and the prison, are emblematic figures" (p. 27). Just as the colony was the "nocturnal face" of the empire, the camp today is the third place where abject precarious life is maintained. This violence is spread far and wide, and now "a new kind of war, utterly planetary, has apparently already been launched and is unfolding on all fronts, being imposed upon us entirely from the outside" (p. 30). This war is fueled by viral affect and fear, destroying any possible relationality that might help diverse peoples trust each other. There is a general atmosphere of fear that "feeds into the idea that the end of humanity—and thus of the world—is near" (p. 31).

Mbembe paints a disturbing picture of our current condition, a time that is "far from being one of reason," but instead a time filled with desire for "mysteries and the return of a spirit of crusade, ours is rather a time of paranoid dispositions, hysterical violence, and procedures to annihilate all those that democracy will have constituted as enemies of state" (p. 41). He suggests that this renewed mytho-religious fervor is simultaneously flourishing alongside algorithmic reason and its tentacular reach into capital (p. 51). The new "digital divine" is seen as "vital, visceral, and energetic," aligned with new forms of populism and despair (p. 51). What emerges is the "security state," which is premised on the insecurity of its virtual hordes, utterly isolated and yet digitally connected, like adjacent foam bubbles, proximal, linked, and yet alone. The emphasis everywhere on machine learning AI that iterates endlessly over a probability landscape of doubt and calibrated credibility, simultaneously and ironically results in large-scale refusals of any kind of healthy doubt, suggests Mbembe, demanding instead a non-verifiable faith in automated intelligent systems (p. 55). This state of affairs rests on and furthers nano-racisms, achieved through control systems operating at the molecular and granular scale of infra-individual affective collectivity. Viral disinformation speeds across the social links of online "platforms," broadcast without apparent limit or correction, whereby belief accrues into relative truth. The emergence of epistemic authority in the midst of this viral affect forces us to think differently today about how thought "spaces itself out into the world."

Mbembe draws from Deleuze's project of "transcendental empiricism," which reads European empiricism as a "philosophy of the imagination" in its emphasis on humans as "inventive species," bound to habit, belief, and passion (Deleuze, 2001, p. 110). Knowledge claims are derived from inclinations, interests, and passions, assembled together into a "general view" that establishes a general idea, or a general interest, an institution, a rule for action, etc. Rather than cast Enlightenment empiricism as purportedly bound to the rational subject, Deleuze helps us appreciate how the imagination was crucial to thinkers such as David Hume: "General interest exists only by means of the imagination, artifice, or the fancy" (Deleuze, 2001, p. 130). Mbembe draws on Deleuze's history of European empiricism, revealing how Enlightenment images of reason emerged during an epistemic crisis regarding the limits of pure deduction and consequent attempts to validate plausible and probabilistic reasoning, alongside the speculative stretch of capitalist venture (wager/debt) through colonial expansion. Deleuze's notion of transcendental empiricism is meant to displace the delusion of the rational subject, while still pursuing an image of reason. Bell (2008) suggests that Deleuze pursues a kind of "infinite pragmatism" in emphasizing the power of the imagination to plug into an actual infinite differential. As Deleuze and Glissant remind us, general concepts and knowledge of the whole (ocean, earth, people, cosmos, etc.) are quivering gestures with a nondenumerable infinite differentiability, and thus find themselves with one foot in the virtual and one foot in the actual. Generalizations and concepts are rules derived from finite habits, but they are achieved only by plugging into an "infinite differential" through the power of the imagination. This infinite undergirding of finite encounters reveals the weakness of current "algorithmic reason," which generates general views based on large amounts of data; finite computational processes privilege identity, similarity, and the repetition of the same or relatively different – and, of course, they suffer from the intractable problem of induction.[2]

Flow and gather

Under these complex conditions, Mbembe turns to Édouard Glissant for a hopeful politics of flow and the passerby. He presents the nomadic passerby as contradictory, a scapegoat for capital, a negligent tourist, and yet also a resistant agent, importing and repurposing the terror of such precarious passing status: "The capacity to move across considerable distances, to entertain shifting alliances, to privilege flows to the detriment of territories, and to negotiate uncertainty is necessary to influence the regional markets of terror" (Mbembe, 2019, p. 37). The passerby is carefully crafted in the work of Glissant to avoid associations that romanticize peripheral participation and diminish the imperfect potentiality of passage. Glissant traces a particular epistemic imaginary across the nomadic wandering of the passerby, delineating between first nomadic laborers whose trajectory is circular, compelled to go and return repeatedly, exhausting and exhausted, and secondly imperial nomads whose arrowlike trajectory conquers and settles land, spreading themselves out and over all borders. Glissant seeks a nomadic movement that might better speak to the epistemic relationality of a "tout-monde" that is not bent on totalitarianism, a rhizomatic network that is neither closed – like the path chased by an abject laborer or the personal adventurer – nor plundering and invasive like the arrowlike nomad: "conquerors are the moving transient root of their people." Through an archipelagic re-imagining of both spatial politics and epistemology, specifically through the fragmenting power of the windward islands, earthdwellers might gather a new way of thinking about residency and belonging, a new mode of relationality. The Caribbean epistemic imaginary puts forward a derangement (Glissant calls this a Baroque derangement) because Relation presents itself most visibly across this archipelago of islands, and because this is one of the "explosive regions" where Relation is "gathering strength." I want to emphasize this term "gathering" in relation to epistemic authority, temporarily moving nomadically with emergent knowledge, thinking with the Caribbean seascape of that word, as it offers radically different ways of practicing epistemology, where gathering seems to better "collect" perspectives, compared to the heavy-handed grasping of "comprehension."

The Caribbean poets of exile that Glissant considers – whose relation to France was the source of their exiled identity – are critiqued for not adequately thinking beyond the notion of center/periphery. These exiled poets continue to long for a dialectical completion, marking leagues between metropolis and distal outpost. Searching for new conceptualizations of a rhizomatic relationality, however, Glissant rejects the language of the electrical circuit and cybernetic information – the circuit of feedback energy: "we are not referring to a circuit, a line of energy curved back on itself" (Glissant, 1997, p. 32). This cybernetic image is still too dependent on the notion of trajectory, a passage that is conscripted and targeted, when "trajectory, even bent or inflected, no longer applies" (Glissant, 1997, p. 32).

From Glissant, Mbembe gleans core lessons concerning space and time: he realizes with a certain sang-froid that our terrestrial future rests on the paradox of "proximate intimacy with those who have been separated" (Mbembe, 2019, p. 44). That is, he suggests we must undergo these conditions within the new configuration of an archipelagic Caribbean Sea, a diffractive sea, as Glissant says, for whom (rather than "in which") indeterminacy and opacity are essential qualities of truth. Another epistemic imaginary is possible only after accepting the uncomfortable proximity of life, that "we are condemned to live in our exposure to one another, sometimes in the same space" (Mbembe, 2019, p. 40). Owing to this structural proximity among postcolonial earth dwellers today, "there is no longer any "outside" that might be opposed to an "inside," no "elsewhere" that might be opposed to a "here," no "closeness" that might be opposed to a "remoteness." One cannot "sanctuarize one's own home by fomenting chaos and death far away, in the homes of others" (Mbembe, 2019, p. 40).

For Glissant, the Caribbean Sea is the perfect place to rethink spatial imaginaries as rhizomatic networks of opaque affiliations that might dethrone the familial line of epistemic patriarchy and its lateral flattening of belonging, which cannot forget the violent journey, but nor can it stay in place, defending the inside from an outside that determines ownership and territory:

> The notion of the rhizome maintains, therefore, the idea of rootedness but challenges that of a totalitarian root. Rhizomatic thought is the principle behind what I call the Poetics of Relation, in which each and every identity is extended through a relationship with the Other.
>
> *(Glissant, 1997, p. 11)*

This "Relation" is the turbulent confluence "whose globality organizes the chaos-monde" (Glissant, 1997, p. 94). Relation must be more than a set of linked nodes segmented through conventional spatial logics (the node/line diagram) and become instead a web or mesh by which tension forms a gathering in a fabric. He emphasizes the "total" nature of Relation, and how the rhizome helps us rethink part-whole relationality:

> To the extent that our consciousness of Relation is total, that is, immediate and focusing directly upon the realizable totality of the world, when we speak of a poetics of Relation, we no longer need to add: relation between what and what? This is why the French word Relation, which functions somewhat like an intransitive verb, could not correspond, for example, to the English term relationship.
>
> *(Glissant, 1997, p. 27)*

A poetics of relation must be different from narratives of communal heritage, for these are dangerous treatises intended to assert primacy of affiliation and origin, and their totalizing is top-down. Glissant critiques Christian narratives about origins (sin, man, spirit), as well as Darwinian "generalization" whereby evolution is taken as a teleological indicator of progress and fitness, in Herbert Spencer's extension of the ideas to "social evolution." He turns to the power of the imagination to rethink totality as flow and gathering. His approach to epistemology and earth-aesthetics supports a new "general view." An imagined poetry must no longer conjecture and affirm a given people, but affirm "the evolution of the planet Earth" (Glissant, 1997, p. 32). He seeks a critical postcolonial earth aesthetics, a new relational onto-epistemology.

Queering space-time

Keeling (2019) opens her book *Queer Times, Black Futures* with a quote from Glissant, who figures throughout the book as a poetic force and source of inspiration. She thinks with Glissant's gathering of an archipelagic "totality" so that we might "live the totality of the world as a temporal layering of the linked histories of peoples" (Keeling, 2019, p. 198). She affirms a surrealist effort to combat "miserabilism" and suggests we pursue the surrealist insistence upon encountering the marvelous, embracing poetry, the absurd, the cosmic, etc. The surrealist position is cited as an awareness that "poetry, not formal knowledge, is 'the most certain guide to action' wherein 'poetry' might be a property of things, rather than simply a human capacity" (p.xiii). Notably, the surrealist André Breton visited Martinique in the 1950s and strongly influenced Aimé Césaire and Glissant.

Following Glissant, Keeling treats poetry as part of an epistemic imaginary, a means to stretch into the future, opening up hypothetical spaces for imagining otherwise and "testing" alternative habits. The risk involved in this is essential, related to the "surplus value" generated by the imagination, and its power to displace current-space time configurations. The imagination breaks with "commonsense" habits and deranges the present moment, queering time and the temporality of knowing and being present. As Keeling (2019) suggests, "queer temporality ... names a dimension of time that produces risk" (p. 19). A more adequate attention to the power of the imagination can help discern the nature of this risk, perhaps pursuing the kind of radical imagination described by Haiven and Khasnabish (2010):

> We approach imagination as a process by which we collectively map "what is," narrate it as the result of "what was", and speculate on what "might be". It is cognitive and corporeal, intensely creative and utterly mundane all at once. And while imagination is a terrain of political struggle it is not merely reducible to "ideology" in any simplistic sense of "false consciousness" or "fetishism". Imagination represents a more rich, complex, agent-driven and ongoing working-out of affinity. It is a crucial aspect of the fundamentally political and always collective (though rarely autonomous) labour of reweaving the social world.
>
> *(p. iii)*

Keeling turns to Afrofuturism as a method for pursuing alternative epistemic imaginaries, seeking freedom projects and tactics for queering time and space, and bringing forth the new; she begins with Sun Ra's *Space is the Place*, an intergalactic perspective on race politics, which propounds the need for an extra-terrestrial empirical practice situated in the cosmic power of art and music. She sees a non-innocent creolization and the imagining of errant futures in science fiction, explored in the film Pumzi (2009) by Wanuri Kahiu and the novel Who Fears Death (2009) by Nnedi Okarafora. The book ends with a section called "World Galaxy" – a reference to a song by Alice Coltraine, and a brief discussion about her "black orientalism" – before revisiting some of Glissant's ideas about the poetics of relation and how his vision might help us achieve a state of antifragility (p. 199). In World galaxy, Keeling challenges us to think at this cosmic scale, with the indeterminate "tout-monde," whereby a re-politicization of ecology involves recognizing the antagonistic relationality in which we are invested, and the need to reshape dependency relations. There are tensions between Keeling's project and the politics of radical refusal because she explores the workings of the "Black radical imagination" as an effectuating intervention that continues to inspire action. In search of those who produce "freedom dreams," but with the kind of wisdom that someone reflecting both on the promise and failure of past such dreams, she notes the need to dwell on the spatiotemporalities of our senses as a way of intervening and interrupting the smooth rationalizations of capital's intrusive methods of control. Keeling advocates for "queer futures," or worldings that depend on our investment in a kind of renewal and creative invention that "safeguards the Diverse."

And yet Keeling (2019) persistently reminds us that "critical political investment in foreseeable futures" has been undermined in so many contemporary contexts, and that "fear and vulnerability" preclude the becoming of other worlds (Keeling, 2019, pp.xii–xiii). She cites Kudwo Eshun's (2003) attempt to distinguish the power of speculative fiction from that of corporate venture capital. Finance capital, for instance, pursues the utility of an imagination, designing "Future scenarios" in order to predict and control. In the 1970s, such scenarios were generated by the Royal Dutch Shell corporation in its effort to imagine future energy needs. Keeling shows how future scenario mapping is a speculative skill serving corporate venture capital, and although utterly antagonistic to Keeling's project, she realizes the importance of situating her work in the midst of this economic activity. Royal Dutch Shell thrived through its operations in Africa; the very existence of this corporation "is predicated on a system of racial capitalism that thrives on the dispossession and exploitation of Black people, indigenous peoples (some of whom describe themselves as 'Black'), and people of color" (Keeling, 2019, p. 5).[3]

Shell's "future scenario team" was interdisciplinary, conceived from the corporate perspective as the best way to get around the disciplinary knowledge that shackled current practices/beliefs about energy or life; these established disciplinary knowledges curtailed Shell's desire to imagine, shape, and control the future. Thus, the epistemic "interdisciplinary" here is a trope that serves the continued proliferation of capital's tentacles, as we shift from Foucault's "disciplinary societies" to Deleuze's "societies of control." Deleuze's essay on control societies (Deleuze, 1992) reminds us that alt-futurisms must take up and interrogate the modulating power of 21st-century capitalism. Keeling (2019) states that the imagination is "among the weapons on offer in societies of control" (p. 13) and that imaginations are weapons that can be used to fight the sly abuses of a society of control, but Keeling reminds us, citing Franco Berardi, the imagination can also simply further the intrusive power of control, through the "aestheticization" of everyday life, or the entrepreneurial prompt to turn each day into a work of art. The imagination is never innocent.

Keeling (2019) notes that risk-seeking and creative entrepreneurial adventurists have always profited from "the queerness that is endogenous to time"; colonial conquest in the 17th century produced a system of credit and the existence of "imaginary value," due in part to the distances and long times entailed in the transatlantic slave trade (a year for a boat to leave and later return) (p. 24). Baucom (2005) showed how the modern system of speculative finance developed through the slave trade, when the imaginary value of "mobile capital" came to dominate economics; the present phase of financialization inherits this habit of economic speculation. Thus, the speculative stretch of the imagination can lead to an epistemic violence that comes through "illegitimate beliefs and absurd general ideas" (Deleuze, 1991, p. 127). Humans are fully capable of imagining that a white powder falling from the clouds, resembling snow, might still taste of salt and burn the flesh, and that trees will flower in December in Scotland. Deleuze states that the most telling sentence in Hume's treatise on empiricism is: "Tis not contrary to Reason to prefer the destruction of the whole world to the scratching of my finger" (Deleuze, 1991, p. 33). Deleuze and Keeling help us understand the destructive force of epistemic imaginaries and the way that the imagination works in relation to reason and power, destructively de-animating and disempowering the future.

I emphasize here Deleuze's powerful claim that *the power of the imagination is to imagine power* (Deleuze, 1991, p. 59); that is, to break power off from the exercise of power, to play with the limits of sense and sensation, liberating forms by both "extending them infinitely" and haphazardly "presenting the accidental as essential" (p. 59). The imagination plugs into an infinite differential, breaking power off from its application, in order to guide us as we grapple with the accidental, or that which is taken as random, or perceived as utterly unexpected, original, out of sorts, new, and incommensurable. Speculative claims stretch across such events, inventing new narratives and glossing them with comforting explanations. A tendency to interpret the accidental as essential, perhaps when a storm bursts into lightning and one imagines the hand of god, reveals how the imagination is often used to tame chance.

New media and new aesthetics

Keeling channels Glissant as a media theorist, attending in particular to epistemic imaginaries within the United States. Keeling also derives much of her theoretical force from Deleuze, his essay on Melville, and his books on cinema, where he recounts how the movement-image was displaced by the time-image in post-war Western cinema, through a discussion of Italian neo-Realism and other developments (Deleuze, 1997, 1989a, 1989b). She draws attention to how film and image serve a particular epistemic imaginary, a way of both seeing and knowing, of fabulating and rendering. Photography, cinema, TikTok, or print – Keeling's perspective supports new materialisms and the fact that new media restructure modes of relationality. The historical point for Deleuze, for instance, is that "the purely optical and sound situation" of post-war cinema took over from the "faltering sensory-motor situation" of early cinema, after the destruction wrought by the violence of WWII, when the Anglo-European belief in the world and the body's willful trajectory was compromised. Keeling draws on this historical account and follows Deleuze's insights in linking pragmatism to 20th-century media perspectivism, when he speaks of the tracking shot and the panoramic shot, two shots which convey stationary process and infinite speed, whereby cinema engages with the moral dilemma as to how to deploy these two irreconcilable spatiotemporal orientations. For the media theorist, these material and technical procedures structure the onto-epistemology of the times, as they shape the spatial logic of the 20th century. The tracking shot follows the individual, while the panoramic empties the space of people and agents. The panoramic shot produces an image of land – open and empty and waiting for arrival. What is autochthonous to that space is "there now" only to serve the conquering perception of empty, open space. Keeling, following Deleuze, suggests this is aligned with the discovery trajectories of the new world, and that such cinematic imaginaries in fact contributed to the ontologies of settler colonialism as well as tendencies toward nationalism and white supremacy. When the representational movement-image crumbles, Deleuze (1989b) suggests that "the interval is set free, the interstice becomes irreducible and stands on its own," thereby initiating the reign of "incommensurables and irrational cuts" (p. 277). Incommensurability becomes the new relation.

Although Deleuze wrote his cinema books prior to the explosion of video – and prior to the rampant broadcasting of the moving image in pocket computers distributed worldwide – Keeling argues that he has noticed something that holds sway in our computational culture. This modulating, self-deforming habit of the cinematic image is also the mark of a control society mechanism. Nearly 40 years since Deleuze's books on cinema were written, Keeling argues that "digital media amplifies these relations" (Keeling, 2019, p. 115). Writing today with these ideas, Keeling declares: "Something remains 'cinematic' within the digital regime: The proliferation of irrational cuts and the autonomy of sound-images form the general economy of images, which are primarily characteristic of the digital regime of the image, mark such continuities with the filmic regime" (p. 114).

For Keeling, this opens up a space for new imaginaries and queer futures. She analyses Afrofuturistic film and science fiction to show how this works. One feels a certain vertigo when the irrational cut forms the dominant mode of conjunction, and yet this is precisely how a means of gathering anti-fragility might emerge. In this way, "the cut" modulates and sustains a certain continuity (like the smoothing act of the speculative imagination, as it tries to make sense of the random), but not the smothering kind that is always commensurable with dominant regimes of thought. Instead, the cut makes the incommensurable into a relation. This is one way to care for that which is fragile, and to manage risks in unpredictable times, as Taleb (2007) suggests, not by smothering their contingency with desperate controlling predictive analytics, but learning how to thrive in conditions of disorder and randomness – or perhaps as Deleuze suggests, we must learn how to affirm all of chance.

Keeling (2019) turns to another text by Deleuze, his essay "Bartleby; or, The Formula," in which he analyses the writing of Hermann Melville and U.S. culture, circa 1850 and 1960, respectively. "Bartleby or the Scrivener: A story of Wall Street" presents a character who says little else but "I would prefer not to," when asked to perform at work on Wall Street. For Deleuze (1997), Bartleby is "mobile passivity" and the fantasy of "an original" associated with American pragmatism. But Bartleby is also a story about the jagged edges of truth:

> It is first of all the affirmation of a world in process, an archipelago. Not even a puzzle, whose pieces when fitted together would constitute a whole, but rather a wall of loose, uncemented stones, where every element has a value in itself but also in relation to

> others: isolated and floating relations, islands and straits, immobile points and sinuous lines—for Truth always has "jagged edges."
>
> *(Deleuze, 1997, p. 86)*

Does the Bartleby story call into question the American dream of manifest destiny and open road, a particularly destructive epistemic imaginary? Keeling (2019) points out that the story was written in 1853, two years after the Indian Appropriation Act, three years after the 1850 Fugitive Slave Act, and eight years after the phrase "Manifest Destiny" was coined. The Bartleby story insists on a kind of radical contingency and diagonal movement, which is not, however, simple negation, but a kind of withdrawal – a withdrawal even from saying no. For Keeling (2019), the refrain of Bartleby, who would "prefer not to," is a formula for refusing capitalism, a kind of "decreation" whereby the subjunctive alternative world opens up diagonally so that "what could have not been but was becomes indistinguishable from what could have been but was not" (p. 185). Keeling will say: "Bartleby's passive mobility, his radical refusal, carries a potential to de-create the American project" (p. 186). According to Keeling, this involves rethinking risk and uncertainty. She cites Taleb (2007, 2012), who argues that current epistemic theories of risk measurement have led us astray, wrongly based on assumptions that our knowledge is potentially complete; instead, we need a measure of fragility, or volatility, so that we might better understand when a situation needs care, or needs to be made more robust (antifragile).

Creolization and porous borders

Glissant also asks that we think with an archipelagic diffractive thought that opens up new spatial imaginaries about relationality. One of his key concepts is "creolization" as a form of "baroque derangement," which messes with any classicism entrenched within a postcolonial culture (Glissant, 1997, pp. 91–127). The baroque is the act of "incessant turnover" and proliferation, undoing the doxa of classicism. This use of "baroque" seems a reference to a key Deleuzian term, and indeed Glissant uses it as a positive term of art, to describe the specific scattering of languages that comes after classicism, the creative "echos-mondes" that diffract. He lists examples:

> William Faulkner's work, Bob Marley's song, the theories of Benoit Mandelbrot, are all échos-monde. Wilfred Lam's painting (flowing together) or that of Roberta Matta (tearing apart); the architecture of Chicago and just as easily the shanty towns of Rio or Caracas; Ezra Pound's Cantos but also the marching of schoolchildren in Soweto are échos-monde.
>
> *(Glissant, 1997, p. 93)*

This heterogeneous list affirms the incommensurable variety of such proliferation, but each performs as a kind of spoken language, some more negative than others, a form of expression that is generative of the new, and yet derivative by way of its echo-status. The baroque derangement confirms yet contests classical modes, acting out an ambivalent repetition which nonetheless forges the new. For Glissant, the Creole language is also a discourse that must repeat the colonial language of the invader, but escapes its bonds through excessive folds and twists: the échos-monde are "whatever, coming from tradition, enters into Relation; whatever, defending a tradition, justifies Relation, provides the basis for another full-sense to Relation; whatever, born of Relation, contradicts and embodies it" (p. 93). Glissant helps us to see how the imagination is not free, but works with habits and entrenched rules for action, rules that are used and broken in the act of imagining otherwise. There is risk in all this, as the baroque will always "relay" classicism, and these repetitions will always tend toward universalisms. In the case of Creole, there is the specificity of the French language, and its movement from the "rational" language of the French metropolis to the country/village French that came to the Caribbean.

Chude-Sokei (2015) argues that creolization, as a "Caribbean intellectual tradition," is a necessary supplement to posthumanism and Afro-futurism (p. 129), allowing the latter to better engage with black diasporic thought (p. 129). He shows how Caribbean dub music merges with speculative science fiction, informing alternative epistemic imaginaries associated with hybridity and cyborgian futures. In particular, he takes up Glissant's discussion of creolization in terms of AI, arguing that AI is a "logical extension of creolization as a black posthuman technopeoetics" (p. 130). He also suggests that creolization poses a threat to "the hegemony of an Anglophone black diaspora predicated on the assumptions generated primarily by American racial thinking" (p. 138). Creolization is inevitable in a world where "contact" between different peoples entails both a politically situated territorial conflict and, importantly for Glissant, a transhistorical process that is utterly indifferent to human interests. Glissant insists that creolization can never be a ground of identity because it "can never be accomplished, nor can we go beyond it." Creolization is a process as broadly conceived as evolution, "or rather a broader evolutionary process without man as its center" (Chude-Sokei, 2015, p. 141).

I am sensitive to concerns that terms like "rhizomatics," "archipelagic," and "creolization" should not simply be absorbed into an all-consuming "global" theorizing, as Mimi Sheller (2003) explains. I am very cognizant of its differentiated historical framing – on certain islands, hybridity is used to bolster an elite – and I am reluctant to cast the space as a ur-creole ultra-other domain of romanticist blending or mixing. Creolization cannot function as the ultimate modulation of Western knowledge practices, a process they undergo to be born again in innocence and alterity. At the same time, another smothering catchall might itself be postcolonial theory which would neglect, as Silvio Torres-Saillant (2006) suggests, the specific thought produced in the Caribbean. Of the Caribbean, Glissant (1997) writes:

> This has always been a place of encounter and connivance and, at the same time, a passageway toward the American continent. Compared to the Mediterranean, which is an inner sea surrounded by lands, a sea that concentrates (in Greek, Hebrew, and Latin antiquity and later in the emergence of Islam, imposing the thought of the One), the Caribbean is, in contrast, a sea that explodes the scattered lands into an arc. A sea that diffracts.
>
> *(p. 33)*

We hear in these words the emphasis on a new spatial logic emerging from the Caribbean and its distinctive flow. Concerns that such a celebration of plastic proliferation fails to reckon with the historical fact of embodiment are taken up by Jackson (2020), who extends

Mbembe's critique of biopolitics, and recounts how black female flesh has historically been imagined as that which cannot hold form, a site of modulation and transmutation (p. 3). She shows how blackened people produced within the legacy of colonialism have been coded as plastic and "infinitely mutable," so that slavery and other modes of subjection might better exploit them. Jackson points to the violence of neoliberal humanism and states that it cannot be reformed through creolization. Blackened people are appropriated and "plasticized in the methodology of abjecting animality," which continues today, she contends, in the turn to epigenetics and its emphasis on relational, situated corrections of past genetic determinisms; according to Jackson (2020), these ecological approaches to the body continue to position black women's bodies as toxic environments. Her salient criticism sustains attention on the unlivable conditions of persistent racisms, while turning, notably, to the power of the imagination and the work of contemporary visual artists who are able to seed alternative epistemic imaginaries. And like Mbembe and Keeling, she turns to Glissant, and attempts to "gather" rather than "grasp" knowledge: "to grasp (comprehend) has a fearsome oppressive meaning" (p. 130). Jackson (2020) advocates for an epistemology that moves as an open whole, citing Glissant:

> In Relation the whole is not the finality of its parts: for multiplicity in totality is totally diversity. Let us say this again, opaquely the idea of totality alone is an obstacle to totality.
>
> *(Glissant, 1997, p. 192)*

In other words, the need to re-imagine part-whole relationality depends on the power of the imagination and its speculative force in thinking totality and diversity, or as Jackson (2020) writes, "Glissant's totality is speculative theory that resists the lure of the essentialisms I have described" (p. 131).

Concluding comments

Insofar as an epistemic imaginary emerges through spatial politics and new media, we are led to critique theoretical frameworks that seek a reconfigured identity under new terrestrial conditions, but fail to adequately problematize the question "who belongs here?" And yet, what might be an adequate response to this question today? How might we learn from Glissant's alternative spatial logic that refuses the center-periphery and affirms an archipelagic relationality (Stephens & Martinez-San Miguel, 2020)? I would like to suggest that there is no good response except one that reckons with the feeling of vertigo (a derangement) felt in the face of indeterminacy when pursuing the power of the imagination to imagine power. I am particularly interested in Glissant's use of the word "vertigo" and "vertiginous variances," and in how this vertigo figures in current epistemic imaginaries mapped by both Mbembe and Keeling, and in how vertigo factors in postcolonial opacity and other spatial sensibilities. This is a vertigo felt in the face of uncertainty and quantum epistemic imaginaries, and also a vertigo induced by the memory of the abyss of the slave boat and the Middle Passage. It is the troubling feeling of this epistemic and ontological vertigo that we must stay with. One of Glissant's powerful insights pertains to the limits of critical projects that cannot live with that vertigo, opting instead for a feeling of absolute certainty. His concern is that re-establishing identity through roots is often a violent act of laying claim to land, settling permanently, and occupying in the name of filiation. Instead, he proposes an errantry that might gather knowledge differently:

> The thought of errantry is not apolitical nor is it inconsistent with the will to identity, which is, after all, nothing other than the search for a freedom within particular surroundings. If it is at variance with territorial intolerance, or the predatory effects of the unique root (which makes processes of identification so difficult today), this is because, in the poetics of Relation, one who is errant (who is no longer traveler, discoverer, or conqueror) strives to know the totality of the world yet already knows he will never accomplish this, and knows that is precisely where the threatened beauty of the world resides . . . . The thinking of errantry conceives of totality but willingly renounces any claims to sum it up or to possess it.
>
> *(Glissant, 1997, p. 20)*

The poet learns from those who whisper in the corners, those darkened spaces where a too-bright light will obliterate the nuance of shadow and sublimate the vertiginous truth of the relative (de Freitas et al., 2022; Quashie, 2021). But Glissant's opacity is not meant to be an obfuscation or watery relativism. This is not (only) the relativity of truth, but must also be the truth of the relative, which gathers together not just an aggregate of particulars, but an opaque tout-monde. The opacity of being, for Glissant, is a speculative practice of "totality gathering strength." As much as epistemic imaginaries are provisional and accidental, they are also a way of engaging with an agentic earth, a cosmic galaxy, a queer future – a gesture that stretches into imagined totalities.

> Here, poetic thought safeguards the particular, since only the totality of truly secure particulars guarantees the energy of Diversity. But in every instance this particular sets about Relation in a completely intransitive manner – relating, that is, with the finally realized totality of all possible particulars.
>
> *(Glissant, 1997, p. 32)*

Gathering this kind of collective credibility is particularly challenging in an environment saturated with the accelerated logics and viral affect of late capitalism. In research that learns from poetry, epistemic credibility is gathered through the political power to "represent" under-represented others, not in the way the nation-state or other forms of governance do, but through a strange system of speculation, experimentation, inference, rhythm, and proof, by which we might convoke an authority, furnish a situated truth claim, and explore the consequences of an epistemic imaginary.

Notes

1 I would like to thank everyone who participated in the Glissant reading group, hosted by the Manifold Lab, Manchester Metropolitan University (2018–2021). In particular, thanks go to Laura Trafi-Prats, Yvonne Sinclair, and Stephanie Daza-Curley.

2 Hume's problem of induction states that no amount of finite instances can furnish adequate evidence for belief in the general case.

3 Shell operations in Nigeria in the 1990s are recounted in Keeling (2019). Ken Saro-Wiwa organized laborers and demanded compensation or the cessation of oil extraction in the region. The government refused, and after a sham trial in 1995, Saro-Wiwa and eight co-defendants were hanged. Although Shell Oil condemned the executions, the company also benefited tremendously by them. In 2009, prior to a trial that was brought against Shell, they awarded the family of Saro-Wiwa US$15.5 million, claiming it was a humanitarian gesture and not a mark of culpability.

References

Baucom, I. (2005). *Specters of the Atlantic: Finance capital, slavery, and the philosophy of history*. Durham, NC: Duke University Press.

Bell, J.A. (2008). *Deleuze's Hume: Philosophy, culture and the Scottish enlightenment*. Edinburgh: Edinburgh University Press.

Chude-Sokei, L. (2015). *The sound of culture: Diaspora and Black poetics*. Middleton, CT: Wesleyan University Press.

de Freitas, E., Trafi-Prats, L., Rousell, D., & Hohti, R. (2022). A poetics of opacity: Towards a new ethics of participation in gallery-based art projects with young people. In L. Trafi-Prats & A. Castro-Varela (Eds.), *Visual participatory arts based research in the city* (pp. 126–142). New York: Routledge.

de Freitas, E., & Truman, S. (2020). *New empiricisms in the anthropocene: Thinking with speculative fiction about the future of social inquiry*. Qualitative Inquiry. https://journals.sagepub.com/doi/full/10.1177/1077800420943643

Deleuze, G. (1989a). *Cinema 1: The movement image*. (Trans. H. Tomlinson & B. Habberjam). Minneapolis, MN: University of Minnesota Press.

Deleuze, G. (1989b). *Cinema 2: The time image*. (Trans. H. Tomlinson & R. Galeta). Minneapolis, MN: University of Minnesota Press.

Deleuze, G. (1991). Empiricism and subjectivity: An essay on Hume's theory of human nature. (Trans. CV. Boundas). New York: Columbia University Press.

Deleuze, G. (1992). Postscript on societies of control. *October*. 59(Winter, 1992), 3–7.

Deleuze, G. (1997). Bartleby or the formula. In D. Smith & M. Greco (Trans.), *Essays critical and clinical* (pp. 68–90). Minneapolis: University of Minnesota Press.

Deleuze, G., & Guattari, F. (1994). *What is philosophy?* (Trans. H. Tomlinson & G. Burchell III). New York: Columbia University Press.

Eshun, K. (2003). Further considerations on Afrofuturism. *CR: The New Centennial Review*, 3(2), 287–302.

Glissant, E. (1997). *Poetics of relation*. (Trans. B. Wing). Ann Arbor, MI: University of Michigan Press.

Glissant, E. (1999). *Caribbean discourse: Selected essays* (Trans. M. Dash). Charlottesville, VA: University of Virginia Press.

Glissant, E. (2019). *The collected poems of Édouard Glissant*. (Trans. J. Humphries, Ed., & M. Manola). Minneapolis: University of Minnesota Press.

Haiven, M., & Khasnabish, A. (2010). What is the radical imagination: A special issue. *Affinities: A Journal of Radical Theory, Culture, and Action*, 4(2), i–xxxviii.

Jackson, Z. I. (2020). *Becoming human: Matter and meaning in an antiblack world*. New York: New York University Press.

Keeling, K. (2019). *Queer times, Black futures*. New York: University of New York Press.

Latour, B. (2004). Why has critique run out of steam? From matters of fact to matters of concern. *Critical Inquiry*, 30(2), 225–248.

Latour, B. (2017). *Facing Gaia: Eight lectures on the new climatic regime*. London: Polity Press.

Mbembe, A. (2019). *Necropolitics*. Durham, NC: Duke University Press.

Quashie, K. (2021). *Black aliveness, or a poetics of being*. Durham, NC: Duke University Press.

Sheller, M. (2003). *Creolization in global culture*. New York: Routledge.
Stengers, I. (2017). *Another science is possible: A manifesto for slow science.* (Trans. S. Muecke). Hoboken, NJ: Wiley Press.
Stephens, M., & Martinez-San Miguel, Y. (2020). *Contemporary archipelagic thinking: Toward new comparative methodologies and disciplinary formations*. Lanham, MD: Rowman & Littlefield.
Taleb, N. (2007). *The Black swan: The impact of the highly improbably*. New York: Random House.
Taleb, N. (2012). *Antifragile: Things that gain from disorder*. New York: Random House.
Torres-Saillant, S. (2006). *An intellectual history of the Caribbean*. New York: Palgrave Macmillan.

14

FRAGILE TEXTS AND MACHINE READERS

Trans/in/dividual reading tactics in a complex technical milieu

Introduction

There is little doubt that generative neural network algorithms like large language models (LLM) are good at predictive aspects of language processing.[1] They scrape information from diverse digital registers and achieve a synthesis akin to reading and writing (Pavlick, 2023). The LLM has emerged as part of our investment in neural network approaches to AI, enhanced by transformer algorithms that incrementally map the syntax of its near-infinite digital resources. As part of their training, these algorithms scour the internet, sweep through texts of dubious authority, and rummage through our language rubble. Generative algorithmic infrastructure allows them to simulate and mimic all that they have read. It may sound glib, but I fear that LLMs are the readers who are best suited to our distraction economy and post-truth conditions. Ironically, they operationalize an algorithmic architecture called an *attention mechanism* that decomposes grammar and lexicon into a scheme of plausible "relevance" measures (Vaswani et al., 2017). These "transformers" embody a kind of neoliberal relativism in their performative attention to variety and viral influence, and their affirmation of internet alternative truths. Some advocates argue that LLMs have achieved an "implicit embodied knowledge" of language, and that they are very effective at simulating a "grounded" reading, but there are all kinds of debates about how they do so and the extent to which their language use can be characterized by traditional linguistic and cognitive theories (Millière, 2023). Critics of LLMs suggest that neural network models will always produce poor examples of AI (Chollet, 2019; Marcus, 2022), being stochastic parrots and inefficient learners, lacking symbolic and logical structure as well as any grounded semantics (Bender et al., 2021). And yet our understanding of how these technical models achieve their language skills is nascent, and it feels far too premature to dismiss their cognitive behavior as utterly shallow.

With these concerns in mind, this chapter responds to the following questions: What is reading all about, as our technical milieu becomes increasingly digital and our reading increasingly automated? How are embodied and transindividual practices associated with human reading both similar to and different from current machine reading models?

DOI: 10.4324/9781003538769-14

This chapter works with the ideas of philosopher Gilbert Simondon (1924–1989), drawing primarily from his 1958 book, *On the Mode of Existence of Technical Objects*. In turning to Simondon, I follow others who suggest that his work has some resonance with various new materialisms and posthumanisms in which technical reality is re-imagined (Bardin, 2021). And I channel Simondon's quasi-organic emphasis on the sensory-motor and affective dimension of our technical reality, and the material-kinetic mode of being human, by which, I argue, one can reveal the monstrous trans-in-dividual nature of reading (Simondon, 2022). This chapter is a response to the sudden and widespread international impact of LLMs across many sectors of society and the reshaping of the technical milieu, raising questions about human cognition, exceptionality, and intelligence. Philosophical inquiry into technology is what we need most today, as generative fabulations by infant-like algorithmic mechanisms further diminish our trust in the authenticity of texts. The mediating force of philosophy should, in the words of Simondon, offer "an edification of a culture" and re-assessment of the human-nature-technology relationship, by attending to "technical thought" more carefully and slowing down the convergence toward technocracy (Simondon, 2017, p. 222).

It feels more necessary than ever to engage in reading philosophy texts that help us rethink digital life and the transindividual nature of being (Braidotti, 2018). This kind of reading is what might better prepare us for the battles yet to come with the mobsters at the liars' banquet, gorging themselves on our personal data (Klein, 2023). Simondon offers a response to technocratic advances in automated reading, an alternative philosophical analysis of stochastic parrots (Bender et al., 2021). "The philosophical effort thus finds itself faced with a unique task to be accomplished: the search for unity among the technical and non-technical modes of thought ..." (Simondon, 2017, p. 225). Genealogical and historical studies of the becoming of technics allow one to find that common ground across these different modes of being. This kind of onto-philosophical analysis is urgent, as we enliven a digital world that simulates truth warranting and cognitive labor (Hui, 2016). Philosophical inquiry into the reading and writing of generative algorithms will help us better adapt to this rapidly changing technical milieu.

The book

On the Mode of Existence of Technical Objects was originally published in French in 1958. A new English translation by Cecile Malaspina and John Rogove was published in 2017 by Univocal Publishing. My copy of this book is falling apart from too much attention and too many readings. The cover paper is a thick textured matte white paper, which holds fingerprints and other stains; the letters of the book's title and author's name are embossed onto the thick paper stock, and one can feel the letter edges as one moves fingers across the book's cover. The last page states:

> This work was composed in Garamond. All materials were printed and bound in March 2017 at Univocal's atelier in Minneapolis, USA. The paper is Accent Opaque. The letterpress cover was printed in Lettra Fluorescent. Both are archival quality and acid-free.

How does any of this material framing of the text matter?

I demolished this book while reading it. The binding cracked, and the pages detached, falling out of order. The front cover bears the trace of dirty grasping hands and the evidence of a messy coffee encounter (Figure 14.1). This book now hangs together precariously, and reading becomes a delicate matter of sustaining the object as a unified whole. My dog Gracie used the book as a pillow. Reading involved a decomposition, revealing the book's frailty and fragile fabric. I feel a certain amount of pride for having prised it open so thoroughly, attentive now to its awkward material construction of 281 "paper-thin" layers of printed words, glued and bound together. Reading has unfastened the pages.

But this book came delicate, as though the publishers intended that readers would participate in and witness its demolition; one surely anticipates the book's potential demise as soon as one touches the finespun gauzy paper cover and flimsy spine; wax paper jackets to protect the object could not resist the force of reading. Looking back at what occurred, one might say that reading is a process of demolishing the object, a process that is also a tribute to the fabricated book, the "ontogenesis" and "individuation" of the book as technical object. On the last pages of this book is a list of other Univocal titles in this same fragile series; I own many of these little gems that barely survive their reading. Indeed, I am extremely fond of them all, each published with the same weak binding and the soft matte white of Accent Opaque paper. Univocal puts these precious missives out into the world as though asking that we care for them, that we take responsibility for our heavy-handed human reading. On my shelf, I find other such books which have undergone similar impassioned reading, their pages creased, their binding broken, and their soft white covers smudged and darkened. As with the Simondon book, I have scrawled my thoughts in the margins, insisting that I occupy these pages along with the author (Figure 14.2).[2]

To what extent is this reading practice similar to the practice of LLMs as they "read" the internet, seeking patterns in data and scraping texts for meaning? Am I operating according to an "attention mechanism" by which I sort and reassemble relevant text? How does

FIGURE 14.1 Book undergoes reading.

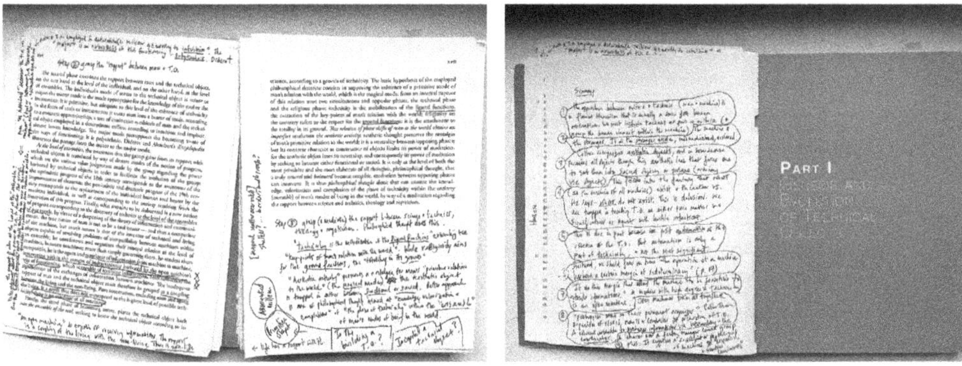

FIGURE 14.2 Interventions in the text.

my method of reading resemble the "transformer algorithm" – which is the engine of the LLM – as it artfully reads words simultaneously, linking each word's relevance to the others, in a vast computational network of recalibrated synthesis? Am I like an LLM, operationalizing hundreds of billions of parameters, roaming over an undulating hypothesis space where "meaning" is correlated with minimum error? I confess to seeking out some key words and highlighting them, and scanning the page or paragraph again, in search of links to other nuanced words, of scribbling the name of another "relevant" author in the margin, of summarizing the argument, etc. But does the embodied material practice of reading amount to nothing else? Am I nothing but a fleshy plodding attention mechanism? Does the LLM mirror my own reading habits? Is the difference merely in speed and scale of resource access? Is it the case that neural network architectures – organic or artificial – are the driving force of reading comprehension? What other structural, material, and meaning-making processes are at work? Simondon's book helps us link human and machine reading habits, as we grapple with this new technical milieu of neural network AI models. Why is this book here? Because we are in the midst of an apparent frenzy of technical activity, in which the project of reading and writing has been properly appropriated by machinic operations and neural network algorithms. As Galloway (2021) suggests, new technical beings "read" in new ways: "a new model of reading will have to be explored, one that is not simply hermeneutic in nature, but also about cybernetic parsing, scanning, rearranging, filtering and interpolating" (p. 290).

Technical milieu

Despite the tattered conditions of my *copy* of Simondon's book, it hangs together beautifully as a brilliant study of the relationship between technics and world. Simondon's ideas influenced Deleuze tremendously – the key concepts of virtuality, singularity, and minor/major, among others, are firmly linked to the work of Simondon (Bowden, 2012). To help us understand his book-as-project, Simondon offers three preliminary sections: a summary, a prospectus, and an introduction, each with a different emphasis and function. The summary states that the technical object has traditionally been treated as merely an instrument of work or consumption or a means of economic reality, and as such, it is in a profound state of alienation. In other words, technical being is always treated as that which dis/serves human being and is never granted any real being in itself. He is critical of tendencies to declaim the artificial/technical as the enemy of the human, for these fail to comprehend the nature of technical existence (Bardin, 2015). We fail to reckon with our earthly environment if we only react with fear of alienation when faced with a technical automaticity that mimics human outputs (Barthélémy, 2015). Simondon promises that his book will correct this, offering a philosophical project that will demonstrate how to expose the "essence" of technical reality, in which the artificial and the natural are imbricated. He lays out the method for delivering on what is promised, a method that involves three stages of philosophical work: the first seeks to grasp the genesis of technical objects, delineating their abstract elements, their individuation and serrated evolution, and their synergistic convergence into ensembles and systems; the second seeks to expose the complex rapport between humans and technical objects, recognizing how these together commingle in the associated milieu of the social-material; the third extends the analysis of technicity into larger worlding processes engaged in shaping our technical culture and "technical reality." Accordingly, he separates the book into three parts: (1) the genesis and evolution of technical objects, (2) man and the technical object, and (3) the essence of technicity.

My scrawling interventions in the margins of his book are an avid attempt to follow his staging, to understand his aims so that I might better follow the complex argument in the coming pages, and in some sense, my reading is also destructive – my aim is to *disentangle and decompose meaning*. It is this gesture of destruction, or perhaps demolition is the better word, that is so rarely acknowledged in the act of reading. If I don't mark up the text and leave some trace of my reading, or, as in this case, make some rather intrusive and extravagant transcription, I feel myself unmarked by the encounter. This is an encounter that is as much pre-perceptual – and quantum – as it is about individual traces and author intentions. Simondon (2020) asks that we consider the power of "quantum theory" to help us think the complex ontological mixture of human and technics (p. 136). We might then characterize Simondon as a kind of quantum materialist, keen to find a technicity beneath Newtonian and Cartesian mechanistic philosophy (Bontems & De Ronde, 2019). Reading would then be considered both *transdividual* (quantum, infra-individual, intra-active) and *transindividual* (collective, ensemble, network). Thus, in terms of trans/dividuation, reading alters me at the quantum scale, the scale of the infra-molecular, where I dwell in absolute ambivalence, where multiplicity shreds the solace of the subject, a scale of mattering/meaning that offers a profound distributed realism. This makes reading quantum and chance-inflected, a reading beneath the scale of the discerning conscious self who is said to read a text and actively weave together two separate but entangled voices. Reading is transdividual before being transindividual. But how does the human body perform such quantum reading, attending to pre-conscious perceptions of the body, the autonomic sympathetic system reassembling percepts from the page, transducing meaning within the milieu, according to a complex stochastic model? Does such a perspective on human reading practices make human readers somewhat more automated?

Consider the technical discrete "actants" of reading: the word, sentence, paragraph, page. Each of these units is encountered and felt as aesthetic and motor-sensory encounters. Reading the word sparks associations and smells, while the sentence deviates and snakes across the page, the paragraph coheres into a gut feeling, and the book becomes a precious vessel in which one can travel. As we read, the paper edges of cherished books become frayed. Our breath moves in and out with each sentence, and we dwell on the enigmatic echo of a word, taste it, and then swallow the implications of a paragraph. These provisional units of encounter (word, sentence, paragraph) coalesce into a synesthetic whole, by which the reader surrenders self to something that is not simply the aggregate of these sensations; we identify less with ideas, characters, arguments than with the overall sensory milieu of the book. As such, reading seems to help us plug into *both a trans-dividual and trans-individual collective material ecology*, so that the experience of reading is fundamentally unscripted and impersonal, despite all appearances.

Machine readers

LLMs) are a kind of generative neural network model of AI, which have proven impressive at language performance. They scrape information from diverse digital registers and are able to synthesize and compose texts as though they were reading. In one sense, the evolution of machinic or artificial readers lends support to the view that reading is a highly embodied and perceptual process, a matter of roaming over signals, in search of patterns, rather than a matter of innate grammatical structures assumed to be ready at hand. In other words, the associated epistemology of the LLM is said to be *empiricist*, despite the fact that these algorithms perform "immaterial labor" without organic bodies. We might call them empiricists because they diverge from traditional symbolic approaches to AI (Buckner, 2019; de Freitas, in press). Their topological surfing of the darkly folded terrain of plausibility makes neural network models opaque to humans who wonder at their perspicacious capacity. When we ask a LLM to read and synthesize Simondon's seminal book, in the voice of an aspiring philosophy student, it does a decent job, just by brute empirical encountering of other texts. But the fact that such a reading is performed "in the voice of" is perhaps the more disturbing, reminding us how "voice" is largely a simulation. The challenge of reading and synthesizing a difficult text has always involved some manner of fabulation, even as one strives for an accurate rendering. My reading of Simondon is idiosyncratic; my synthesis is always more than a summary, involving creative composition, desire, and error. Deleuze said his reading of Western philosophy made the authors give birth to what they might have considered to be monsters. In like fashion, I scribbled comments into the book margins, broke its binding, and then forced it open to revisit these comments and use them to compose my synthesis.

LLMs use the units of language in different ways than humans, being less organic and differently embodied – they are digital, binary, and recursive loops. The extent to which they encode discrete constituent symbols and rules, organized within predicate-argument structures, is as yet unknown (Pavlick, 2023). Methods for identifying internal structures and causal inferential patterns within the language practices of LLM suggest we might yet crack open the "black box" of their language use (Geiger et al., 2021; Meng et al., 2022). But critics argue that such AI models ultimately fail at systematic compositionality – whereby components can be reassembled so that meaning is preserved – because this requirement depends on the notion of discrete concepts, while LLM models often entangle concepts (Mitchell et al., 2021). In other words, they are not good at separating and discretizing concepts (square, round, red) because they read too widely, indeed they must read obscene amounts of text in order to generate even the most minimal "understanding" of one text, and they struggle to assimilate the many texts they encounter in efficient and coherent ways. Ironically, perhaps this makes current machine readers closer to humans, who tend to operate according to irrational associations and fuzzy concepts, and whose "attention mechanisms" are spread too thin in a distraction economy. And yet humans create concepts and generalizations that seem greater than the combinatoric aggregate, aspiring to a "whole" that seems to entail an instantaneous infinite survey (Glissant, 1997; Ruyer, 2016), and they persistently diverge from classical probabilistic reasoning, as is evident from decades of decision theory research (de Freitas & Sinclair, 2018).

The advocates for LLMs treat language as a stochastic being, something that is inherently statistical in distribution. The model searches for the weighted statistical significance of a text, tracing its citations and echoes, imagining its web of influence (Millière, 2022). Chater (2023) claims that LLM language competency cannot be robust if it lacks a direct perceptual encounter with the "real," by which he surely means an encounter with wet-organic-matter, and yet those who support a Wittgensteinian approach that language use determines meaning might redefine "use" and claim that LLM are "as good a member of the causal chain as anyone" (Pavlick, 2023, p. 10). Like humans who learn through mimicry, these models track the residue that remains from human reading and use this residue to parrot and simulate human text. But they cannot compute the undigitized marginalia, nor the tempo and breath of the diverse readings of any one text. They reduce text to a passive, quiescent, and accommodating object; the LLM finds use-value everywhere, and it never encounters a text outside of its purpose, never a question it won't answer.

Secretive texts

When texts speak, they become ciphers. They express, signal, signify, and propagate waves of signs, as though offering up a smoke signal that slowly becomes sensible across a terrain. These quanta of meaning "make sense" to others when those others resonate with collective activity. The truth is that all books are ciphers. Secret ciphers slink about in all expression. When we read, we sense the secret of the text, not simply the secret of an absent author or a mystery that unfolds in plot, but also the secret of meaning itself. In reading the signal and listening to the object that speaks, we simultaneously resonate with the encoded language and yet also sense a much larger lurking unknown, the continuum of time, and the infinite potentiality of the once-silent page. Reading thus stretches symbolic code and, in so doing, the reader reckons with the hypothetical of the text (the utter silence and noise of a world that does not speak to us). This is a deeply disturbing experience, and so we are inclined to find ways to hear only the message and not the silence or the noise. Reading is an experience in which we are reminded of both the bonds between discrete individuals (the flesh, the code, history) and also reminded of the incommunicative continuum (the imagination, noise, future, silence).

All voracious human readers have their egos stroked by the sycophantic and pleading address of a text – the text invokes, convokes, and provokes, in the spirit of vocare ("to call"). The intimacy of reading is thus never innocent. There is no purity in the encounter. But neither is reading only a matter of situated perspectives and the positioning of subjects in communities bound together through identity. Reading dissipates identity, as much as it affirms it. The *reader-text-creature* is no-one, no-thing, a creature that does not live according to identity and who(le)ness. This is not the critical reader perched and ready for judgment, not even the suspicious reader who has been burned before. This is not the reader who reflects on their biases and reveals their truth. There is a more profound transdividual posthuman work going on when one reads and studies a text. And perhaps for that reason, reading closely and studying a text might be a process of approaching the irresolvable multiplicity of existence, because it is an occasion when we are somewhat more generous, more delicate, more trusting, and less negligent than usual. I have learned over the years that one must honor the text, raise it up as an object of significance, turn it over in your hands, and then disappear inside it, to palpate the semantic envelope from the inside, and then awkwardly resurface, if you dare. This is a process that requires superhuman strength; to become no-one and every-one offers a dangerous euphoria and yet also a way of being present that comes with a different kind of responsibility. The bliss of reading emerges in this monstrous mixture when texts are wrenched from their resting place and readers are unrecognizable to themselves. We can think of this in terms of cataclysmic and mutant transformations that reassemble matter and meaning. The LLM, on the other hand, never encounters a difficult text, never confronts an unknown word, and never fails to recognize itself in its delusion of distributed relevance.

Deep learning

Notably, the publication of *On the Mode of Existence of Technical Objects* coincided with the 1958 launch of the artificial neural net, by Cornell psychologist Frank Rosenblatt who created the "perceptron" machine, a 5-ton computer that taught itself how to distinguish punch cards that were marked on the right or the left, through sampling and self-correcting. A *New York Times* headline stated: "New Navy device learns by doing: Psychologist shows embryo of computer designed to read and grow wiser" (1958). Current machine learning models are derived from this earlier attempt to perform a kind of automated *reading*. This connectionist neural network approach to AI was displaced for decades by efforts in symbolic reasoning, but since the 2000s, layered neural networks have returned to dominance in today's AI models, shaping the current LLMs, which operate with distinctive "transformer" structures and other clever mechanisms. These models can only operate with large-scale datasets, and they are hugely inefficient learners compared to humans. These deep learning models roam across a vast data landscape, a knowledge-milieu of accumulated stories and texts, mobilizing a calculus of gradient descent, which allows them to minimize error as they explore the nature of emergent meaning. These "creative" machine readers are capable of artful expression and simulation and *deep fakes*, just as humans are also capable of parroting, mirroring, copying, and forgery (de Freitas, 2022). In the 1950s, Rosenblatt was trying to build a machine that behaved like other neural networks in humans and animals. Clunky and limited in the way chance was concretized in its shifting probabilistic weights, the connectionist paradigm has evolved into neural network infrastructures that are now extremely impressive in their ability to simulate human language use. The neural network technicity of the perceptron persists in these newer generative models, but access to excessive amounts of computational power and data has changed the technical milieu (Buckner, 2019).

Simondon's attention to technicity and the "ontogenesis" of technical objects was composed in the 1950s and was offered as a contrast to cybernetic mechanical materialisms and formal "molding" idealisms (De Boever et al., 2012). He was critical of Marxist philosophy for its lack of historical *ontogenetic* analysis, and its failure to adequately theorize agentic matter (Bardin, 2021). Voss (2019) critiques Simondon for not adequately attending to the economic forces that shape developments in technology, and she is concerned that people use Simondon to "celebrate the apparently self-sufficient, intrinsic aspect of technological functioning" (Voss, 2019, p. 5). The challenge is to think with Simondon at multiple scales, exploring the technicity of the element (the neural network in this case) while also exploring the socio-technical reality that emerges, a challenge that he himself meets by integrating biological analyzes of auto-kinesis with cross-cultural studies of religion and cultural formation. To follow Simondon's philosophical aims today would be to study the *ontogenesis and technicity* of the artificial neural network as a technical object that emerged in the 1950s and has come to shape our current investment in machine learning as part of a cultural paradigm. To answer Voss's (2019) concern is to extend this study, revealing the consequence at individual and transindividual scales, revealing how algorithmic technicity is linked to current material practices (i.e. human reading practices).

Power in numbers

My argument here is that both human and machine reading involve a kind of distributed mind and networked body and gestures toward the transindividual movement of thought, which recruits at the quantum or infra-scale and scales up across bodies and groups. There is pleasure and rigor in this collective labor, a paradoxical mix of attention and suspension, of care and despair. Over the last decade, I have co-organized reading groups that have been intensive and focused, both in-person and online, and have brought strangers and colleagues together, with no other objective than to collectively share in reading and thinking with texts *as a group*. In 2017, Simondon was selected for group study. At group meetings, I insisted that we "stay close to the text" and try to understand how "the book is the author's project," while also an "objectile" with its own path, an agent of change operating through word, sentence, and paragraph. In the group, we try to stay tuned to how books come into being at particular historical times and places. Much like the LLM, our collective reading is a process of releasing the text from its original mooring, allowing it to travel away from the historical moment of composition. After slow and intensive reading, we begin to feel the presence of the author in our midst, but also the spreading of this presence across past and future.

There is an ethical and political commitment in collectively reading a text, to situating thought within a history of ideas and people, and to reckoning with the material constraints that curtail the quiet, preparatory, and solitary work of reading and writing in a distraction economy. Members of a reading group must set aside obligations and other demands and come together to meet with no other purpose than to read together and study a text, without knowing whether such reading will supply them with any exchange value. The reading group is a way to push back at economies of value that cannot accommodate studious scholarship, to steal back time from academic institutions where reading is sacrificed in the name of productivity. Reading in this way is a slow intellectual endeavor, at odds with the accelerated encyclopedic tendencies of LLMs.

Most of my reading groups have been centered on philosophy texts, because such texts are all too often considered extraneous "primary sources" out of reach and unnecessary for the development of scholarship within the field of social science research. Who has the luxury of time to spend hours and hours dwelling in difficult paragraphs written centuries ago? This commitment to study a text and show-up ready to discuss it becomes a powerful life force within university departments, a mode of resistance against the persistent demands of productivity. Such groups become the sole space where readers are able to create an atmosphere of intellectual inquiry. Such extravagant inefficiency is now frowned upon within the academy.

Knowing that others are simultaneously reading the book makes for a distributed practice and shared solidarity in the challenge. People come to the reading group, ready to reveal the pent-up emotions produced in the solitary reading process. Some are made bored by the book, others angry and frustrated, others elated and transformed; some come only with "this reminds me of …" and are unable to submit to the book (only glancing off it), while others are trapped in the weeds on a particular page. But this act of coming together is a political act of solidarity, and we enter the room as though champions returning home. Champions because we have resisted the corporate university and collectively studied a difficult book of philosophy but also perhaps a little exultant for having demolished the text and become monstrous: the reader-text-creature.

Conclusion

Whenever Simondon writes about transindividuality, it is not confined to the human species but is a field of relations and psychical, biological, and technical processes (Combes, 2013). Simondon's historical ontology attends not just to human processes but also to organic and physical processes by which complex technical ensembles come and go. The challenge is to understand exactly how the human-in-the-loop is situated and agentic in this technical reality (de Freitas, 2021). We humans in the loop need to study how "attention mechanisms" in generative transformer algorithms – one example being LLMs – are both similar and different to human readers. This kind of inquiry cannot be the work only of technicians (at one scale) and sociologists (at another scale), says Simondon, for only philosophy does the hard work of conjoining vastly different scales of mattering. A philosophy of technology must grapple with the unscripted ontogenesis and inventive nature of human technical activity and discern its relevance to material and cultural practices at different scales. Such work cannot be axiomatized:

> It could be that ontogenesis is not able to be axiomatized, which would explain the existence of philosophical thought as perpetually marginal with respect to all the other studies, since philosophical thought is what is driven by the implicit or explicit research of ontogenesis in all orders of reality.
>
> *(Simondon, 2020, p. 256)*

Although his work sometimes fuels proposals for a "transcendental technicity" (Hörl, 2015, p. 9), which may seem to abandon the human, Simondon is hopeful that philosophy will find a way to study the human in the loop. In the midst of strong socio-economic forces that often shape invention, Simondon seeks a "reciprocity of exchange" and even a "social relation of sorts" between the human and technology. The task of philosophy is then to trace technicity across element and ensemble, to study the socio-technical ecology as it evolves, and to cultivate a new technological culture (Simondon, 2011, 2012).

If we are to make sense of LLM reading habits, we need to analyze the technical ensemble (platforms, software, public hype) in terms of the underlying technicity and in terms of our current investment in connectionist computing. We neglect the scale of the technical element (the algorithm, for instance) at our peril, distracted by the fumbling software assemblage and its inadequate sense-making, riven by fear that machines will replace us. In other words, we are distracted by and in thrall to the automation of reading, the simulation of writing, and the learning machine, as that which resembles and replaces the human. And yet, Simondon and more recent commentators focused explicitly on LLM (Chater, 2023) note that machines are not operating in human-like fashion, as far as we know, and that our anxiety about their replacing our labor is not about an essential resemblance between human and machine, but about their output or performance. As usual, the humans are distracted by the product, when it is the process that matters. This anxiety occurs because of our flawed way of individuating, our limited understanding of information and entanglement, and our investment in a kind of technical individualism at the wrong scale of technicity (de Beistegui, 2012).

If we continue to build larger and larger neural networks, consuming more and more computing power, scraping more and more websites, and requiring more and more lithium, this will surely lead to what Simondon calls maladaptation ("désadaption"). Technical maladaptation leads to ecological collapse, as forms of life obsessively iterate one form of technical expression and no other. Imagine a population, like we have today, fixated on screens and keyboards, without any other material skills. Digital life is all-encompassing. To avoid such hypertelec technical fetishism, one must understand the ontogenesis of the undergirding elemental technicity of the digital, and then seek out an alternative inventive open space for the technology, whereby the environment itself is brought forth as responsibly agentic, rather than scraped for its resources. This would be a reading machine that does not entail deadly mining of the milieu, extracting lithium and rare earth minerals to serve our digital desires; this would be a reading process of birthing a technicity that exists virtuality in the milieu and in reciprocity with humans (Simondon, 2017, p. 58). We need to find a way to recognize that *living being* entails an "artificialist way of thinking concerned with constructing automata" but also an ever shifting incomputable (Simondon, 2017, p. 105).

Notes

1 LLM include, for instance, ChatGPT, where GPT stands for Generative Pretrained Transformer. I will use "large language models" throughout this chapter to cover these kinds of technologies.
2 At the time of writing, these other Univocal books are scattered on my shelves: *Biogea* by Michel Serres, *Cosmic Pessimism* by Eugene Thacker, *The Intelligence of the Machine* by Jean Epstein, *Science Fiction and Extro-Science Fiction* by Quentin Meillassoux, *Women Who Make a Fuss* by Isabelle Stengers and Vinciane Despret, and *Cannibal Metaphysics* by Eduardo Viveiros de Castro.

References

Bardin, A. (2015). *Epistemology and political philosophy in Gilbert Simondon: Individuation, technics, social systems*. Dordrecht: Springer.
Bardin, A. (2021). Simondon contra new materialism: Political anthropology reloaded. *Theory, Culture, and Society*, 38(5), 25–44. https://doi.org/10.1177/02632764211012047

Barthélémy, J.-H. (2015). *Life and technology: An inquiry into and beyond Simondon*. Meson Press. Lüneburg: University of Lüneburg Press.
Bender, E. M., Gebru, T., McMillan-Major, A., & Shmitchell, S. (2021). On the dangers of stochastic parrots: Can language models be too big? *FAccT '21: Proceedings of the 2021 ACM Conference on Fairness, Accountability, and Transparency* (pp. 610–623). New York: ACM. https://doi.org/10.1145/3442188.3445922
Bontems, V., & De Ronde, C. (2019) Simondon and quantum mechanics (or, on how the 'preindividual' hypothesis leads to a realistic but non-substantialist interpretation of the orthodox quantum formalism). *Philosophy Today,* 63(3), 611–624. https://doi.org/10.5840/philtoday20191030284
Bowden, S. (2012). Gilles Deleuze, a reader of *Gilbert Simondon*. In A. De Boever, A. Murray, J. Roffe, A. Woodward (Eds.), Gilbert Simondon*: Being and technology* (pp. 135–153). Edinburgh: Edinburgh University Press.
Braidotti, R. (2018). A theoretical framework for the critical posthumanities. *Theory, Culture and Society*, 36(6), 31–61.
Buckner, C. (2019). Deep learning: A philosophical introduction. *Philosophy Compass,* 14(10), e12625. https://doi.org/10.1111/phc3.12625
Chater, N. (2023). How could we make a social robot? A virtual bargaining approach. *Philosophical Transactions of the Royal Society, A*, 381, 20220047.
Chollet, F. (2019). On the Measure of Intelligence. arxiv.org. https://arxiv.org/abs/1911.01547.
Combes, M. (2013). *Gilbert Simondon and the philosophy of the transindividual*. (Trans. T. LaMarre). Cambridge, MA: The MIT Press.
De Beistegui, M. (2012). Science and ontology: From Merleau-Ponty's 'Reduction' to Simondon's 'Transduction'. In A. De Boever, A., Murray, J. Roffe, A. Woodward (Eds.), *Gilbert Simondon: Being and technology* (pp. 154–175). Edinburgh: Edinburgh University Press.
De Boever, A., Murray, A., Roffe, J., & Woodward, A. (Eds.). (2012). *Gilbert Simondon: Being and technology*. Edinburgh: Edinburgh University Press.
de Freitas, E. (2021). Machine speculation/creative abstraction. Invited panel on creativity in research. *International Congress of Qualitative Inquiry,* Urbana Champagne, US, May.
de Freitas, E. (2022). Creative abstraction: The powers of the false in algorithmic reason. Paper presented at the *Annual Meeting of the Society for Social Studies of Science (4s)*, Cholula, Mexico. Dec 7–10.
de Freitas, E., & Sinclair, N. (2018). The quantum mind: Alternative ways of reasoning with uncertainty. *Canadian Journal of Science, Mathematics and Technology Education,* 18(3), 271–283.
Galloway, A. R. (2021). *The uncomputable: Play and politics in the long digital age*. New York: Verso Books.
Geiger, A., Lu, H., Icard, T., Potts, C. (2021). Causal abstractions of neural networks. *Advancements in Neural Information Processing Sysems*, 34, 9574–9586. https://arxiv.org/abs/2106.02997.
Glissant, E. (1997). *Poetics of relation*. (Trans. Betsy Wing). Anne Arbor, MI: University of Michigan Press. (Original work published 1990).
Hörl, E. (2015). The technological condition. (Trans. A. Enns). *Parrhesia* 22, 1–15.
Hui, Y. (2016). *On the existence of digital objects*. Minneapolis, MN: University of Minnesota Press.
Klein, N. (2023). AI machines aren't hallucinating. But their makers are. *The Guardian*. Available at https://www.theguardian.com/commentisfree/2023/may/08/ai-machines-hallucinating-naomi-klein
Marcus, G. (2022). How to build AI we can trust. The Michael Shirmer show # 292. Available at https://www.youtube.com/watch?v=z8lVk5fCvxk
Meng, K., Bau, D., Andonian A., & Belinkov, Y. (2022). Locating and editing factual knowledge in ChatGPT. Preprint. https://arxiv.org/abs/2202.05262.
Millière, R. (2022). Deep learning and synthetic media. *Synthese* 200 (3), 231. https://doi.org/10.1007/s11229-022-03739-2
Millière, R. (2023). Do language models tell us anything about language learning? Paper presented at *Learning under algorithmic conditions: Contingency, control, creativity. Conference at Adelphi University's Manhattan Institute for Studies of STEM and the Imagination*, April 18, 2023.

Mitchell, M., Lake, B., Fedorenko, E. (2021). When machines speak: Language processing in computers and humans. Panel presentation. Columbia University. Available at https://raphaelmilliere.com/events_list/when-machines-speak/

Pavlick, E. (2023). Symbols and grounding in large language models. *Philosophical Transactions of the Royal Society, A,* A.381, 20220041.20220041. https://doi.org/10.1098/rsta.2022.0041.

Simondon, G. (2011). The limits of human progress. (Trans. S. Cubitt). *Cultural Politics*, 6(2), 229–236. (Original work published 1959).

Simondon, G. (2012). Technical mentality. In A. De Boever, A. Murray, J. Roffe, & A. Woodward (Eds.), *Gilbert Simondon: Being and technology* (pp. 1–15). Edinburgh: Edinburgh University Press.

Simondon, G. (2017). *On the mode of existence of technical objects*. (Trans. C. Malaspina & J. Rogove). Minneapolis, MN: Univocal. (Original work published 1958).

Simondon, G. (2020). *Individuation in light of notions of form and information*. (Trans. T. Adkins). Minneapolis: University of Minnesota Press. (Original work published 1964).

Simondon, G. (2022). *Imagination and invention*. (Trans. J. Hughes & C. Wall-Romana). Minneapolis: University of Minnesota Press. (Original work published 1965).

Vaswani, A., Shazeer, N., Parmar, N., Uszkoreit, J., Jones, L., Gomez, A. N., Kaiser, L., & Polosukhin, I. (2017). Attention is all you need. *31st Conference on Neural Information Processing Systems*, Long Beach, CA, USA.

Voss, D. (2019). Invention and capture: A critique of Simondon. *Culture, Theory and Critique*. https://doi.org/10.1080/14735784.2019.1679652

INDEX

Note: *Italic* page numbers refer to figures and page numbers followed by "n" denote endnotes.

For Product Safety Concerns and Information please contact our EU
representative GPSR@taylorandfrancis.com
Taylor & Francis Verlag GmbH, Kaufingerstraße 24, 80331 München, Germany

www.ingramcontent.com/pod-product-compliance
Ingram Content Group UK Ltd.
Pitfield, Milton Keynes, MK11 3LW, UK
UKHW050229200926
475702UK00020B/2941
* 9 7 8 1 0 3 2 8 7 7 2 0 4 *